Beyond Biofatalism

Beyond Biofatalism

HUMAN NATURE FOR AN
EVOLVING WORLD

Gillian Barker

Columbia University Press
New York

Columbia University Press
Publishers Since 1893
New York   Chichester, West Sussex
cup.columbia.edu
Copyright © 2015 Columbia University Press
All rights reserved

Library of Congress Cataloging-in-Publication Data

Barker, Gillian.
  Beyond biofatalism : human nature for an evolving world / Gillian Barker.
     pages cm
  Includes bibliographical references and index.
  ISBN 978-0-231-17188-5 (cloth : alk. paper) — ISBN 978-0-231-54039-1 (ebook)
  1. Evolutionary psychology. 2. Human evolution. 3. Behavior evolution.
4. Social evolution.   I. Title.
  BF698.95.B355 2015
  155.7—dc23
                                                                       2015012299

Columbia University Press books are printed on permanent
and durable acid-free paper.
This book is printed on paper with recycled content.
Printed in the United States of America
c 10 9 8 7 6 5 4 3 2 1

COVER IMAGE: © Getty Images / Hiroshi Watanabe

COVER DESIGN: © Diane Luger

References to websites (URLs) were accurate at the time of writing. Neither the author nor Columbia University Press is responsible for URLs that may have expired or changed since the manuscript was prepared.

To the memory of Roger and Louise Barker—
Humane scientists, beloved teachers,
practical thinkers, quiet revolutionaries

Contents

Preface ix

[1]
Human Nature and the Limits of Human Possibility 1

[2]
The Cost of Change 21

[3]
Thinking About Change and Stability in Living Systems 35

[4]
Lessons from Development, Ecology,
and Evolutionary Biology 53

[5]
Human Possibilities 67

[6]
Valuing Change 85

[7]
Choosing Environments 97

[8]
What Is Feasible? 115

[9]
Evolutionary Psychology and Human Possibilities 135

Notes 141
References 143
Index 155

Preface

I HAVE WATCHED WITH HOPEFUL FASCINATION the growing interest of social scientists and a larger public in applying evolutionary thinking to human behavior. Our need to understand the roots of human choices and social patterns has never been more pressing. Climate science, ecology, and the other sciences that examine human impacts on the Earth—and on its capacity to sustain us—have demonstrated that the course that the human species is now traveling is a disastrous one. Indeed, I believe that they show that human survival, especially peaceful survival with a good quality of life, requires some fundamental changes in our patterns of behavior starting as soon as possible. But just what changes are the best ones to pursue, and what are the most effective means for bringing them about? These are ancient and difficult questions, but new tools might help to resolve them. The evolutionary approach has been impressively successful in expanding our grasp in many areas of the life sciences, including medicine and the behavior of other animals, and I have dared to anticipate that evolutionary thinking would open a new avenue to a clearer understanding of how we might begin to make the needed changes. I have been encouraged in this hopeful thought by the energetic entry of first-class evolutionary thinkers like Steven Pinker, David Buss, and Robert Wright into the task of pulling the quite complex and detailed relevant research and analysis together into an overall interpretation of the main implications of human evolutionary history.

I delved into the resulting works of synthesis with increasing dismay. Though they are rich with illuminating insights and intriguing empirical results, the overall interpretations that they offer seemed to converge on all-too-familiar motifs of gender differences and tendencies toward aggression, intolerance, and social competition—conclusions that do not square with my own reading of the basic research and my own reasoning about it. The picture that they present is pessimistic, suggesting that human nature is inflexible enough that substantial change to our social arrangements and patterns of behavior may be out of the question, whereas I see grounds for optimism in many of the same sources. Additional research from related areas of biology, psychology, and philosophy, including some that has been published since the major works of synthesis were written, reinforce my sense that the picture these works present is misleading.

The evolutionary approach to understanding human behavior has been controversial from its inception, and many critical studies of human sociobiology and evolutionary psychology have been published. But these focus for the most part on challenging the approach as a whole. My concern is different—I believe that the evolutionary study of human behavior and society has a vital contribution to make, but its leading thinkers have overlooked some of its key lessons. The missing elements are needed not for what Pinker has called "dubious moral uplift"—a feel-good story about human nature—but to guide us in making effective political and practical choices as we confront the stiff challenges of the twenty-first century. I decided that I needed to investigate carefully how the leading synthesizers supported their claims and reexamine their picture of human nature and human social possibilities, developing a new synthesis that takes account of the most recent research and thinking. The result of my efforts is contained in the pages that follow.

This work was undertaken at the Rotman Institute of Philosophy at the University of Western Ontario, a unique center for research at the intersection of philosophy and the sciences. I am deeply grateful for the Rotman Institute's support for this research, and for the stimulating discussions and thoughtful feedback provided by many of my colleagues and students at Western and elsewhere. I especially thank Philip Kitcher, Nicholas Thompson, Bruce Glymour, Stephen Crowley, O'Neal Buchanan, Boyana Peric, and Graham Bracken, as well as three anonymous reviewers, for insightful comments. Patrick Fitzgerald of Columbia University Press has been the most helpful and enthusiastic editor an author could hope for, and Jonathan Barker's wise advice has improved every page. Any errors that remain are mine alone. Thanks finally, and always, to Dave Pearson for unflagging support and for persistent good humor and good sense.

Beyond Biofatalism

[1]
Human Nature and the Limits of Human Possibility

WHAT KINDS OF SOCIETY are possible for human beings? What changes to our current social arrangements are feasible, and how can they be achieved? These are old questions, but they have recently taken on a new urgency. It has become plain that substantial social change of some kind will inevitably occur over the next few human generations, driven by new conditions of population density, resource scarcity, and climate change—conditions that are themselves in turn strongly influenced by what humans do and how we interact with one another. We face imminent and consequential choices about which social changes to pursue and which to resist, and by what means. At the same time, optimism about social change has been fading. Many commentators have noted that the utopian political and social movements of the twentieth century failed to achieve their lofty goals—some led instead to terrible new forms of oppression—and that even in the most enlightened societies many social problems persist despite public policies designed to counter them. These thinkers sometimes go on to draw a broader conclusion: that the possibilities open to human societies are much more limited than would-be reformers had assumed. But this lesson is too vague to be useful. To respond effectively to coming challenges and present problems, we need to understand the possibilities open to us: the particular constraints that limit our options and the pathways by which particular changes might be achieved.

Some of the more forceful recent discussions of the scope of human possibilities have been informed by ideas from evolutionary psychology, taken in the broad sense: evolutionary studies of human cognition and social behavior, what some proponents call "the new science of human nature." This label links evolutionary psychology to the long tradition that takes a conception of human nature as a crucial measure of what arrangements of human society could and should be pursued. A well-known version of this kind of reasoning is found in philosophical debates in the seventeenth and eighteenth centuries about the "state of nature." Is it (as Jean-Jacques Rousseau argued) peaceful, communal, and egalitarian but capable of corruption by ill-advised innovation? Or is it (as Thomas Hobbes had asserted a century earlier) a state of war, all against all, that only an overarching central power can hold in check? Philosophers no longer appeal to the state of nature, but some evolutionary psychologists argue that evolved human nature sets limits on what is possible for human individuals and for human societies—limits that scientists can discover and specify.

Since close attention was first focused on the evolution of human social behavior in the 1970s, a series of prominent and influential thinkers have argued that evolutionary science has discovered fundamental behavioral dispositions written into the evolved genetic makeup of human beings that limit the social arrangements that are possible for us. These dispositions are not supposed to be absolute—they are subject to cultural and educational influence—but it is claimed that they nonetheless play a profound role in shaping human societies. They solidify certain dominant patterns in human social life such as sharply distinct gender roles, hierarchies of social power, and intergroup violence and intolerance, and they impose limits and costs on efforts to modify these features. These thinkers warn that the discoveries of evolutionary psychology warrant pessimism about efforts to accomplish major reductions in inequality, intergroup conflict, or gender role differences. In light of what evolutionary psychology reveals about human nature, they see such efforts as unlikely to succeed and as morally inadvisable.

The flawed understanding of social possibilities put forward by these thinkers in the name of evolutionary science is what I call "biofatalism": a broad pessimism about the prospects for social change that, while not involving a commitment to genetic determinism, is nonetheless based on a particular set of presumptions about the biological underpinnings of human behavior. It stands in the way of an adequate discussion, both scientific and political, of the social changes that will preserve and enhance the quality of life under the stringent constraints of global warming and other environmental limitations. But such a discussion does require close attention to what we can learn about human psychology from evolutionary science, and it can benefit greatly from some of the genuine insights of evolutionary psychology. The broad aim of this book is to examine and challenge the misleading presumptions that lead to biofatalism and to begin an exploration of what can be learned from an evolutionary psychology freed from those presumptions.

The debate aroused by the claims that many feminist, egalitarian, and peace-oriented social objectives are untenable has been fierce, and some critics argue that evolutionary approaches to the question of human possibilities should be rejected across the board. This book takes a different and more positive approach. It aims to show that a broader consideration of evidence from evolutionary biology and related areas of biology and psychology, and from recent work within mainstream evolutionary psychology itself, supports a different picture of human nature, one that shows us to be more open to some important varieties of social change than the leading synthesizers of evolutionary psychology allege. This new perspective on what evolutionary psychology contributes is opened up by clearing away some conceptual obstructions—examining what the leading voices of evolutionary psychology have said about human possibilities, the reasoning behind their claims, and the tacit assumptions built into that reasoning. When we revise these assumptions in light of what biologists and psychologists have learned over the last few decades across many fields of research, we can see that the evolved strategies expressed in behavior are much more complex and open to a variety of influences than the leading syntheses admit, in part because of the multistranded sensitivity of evolution itself: both behavior and evolution are realized by means of complex organism–environment interaction. Key elements of this interaction include the active plasticity that allows organisms to respond to environmental variation and the many processes by which organisms modify their environments in ways that in turn affect both themselves and others (what some evolutionary theorists have dubbed "niche construction"). Related research from social psychology, developmental biology, and ecology reinforces this expanded view of evolved human nature and fills in its details in certain key areas.

The new conception of human nature and its role in human social life offered here supports the conclusion that some substantial and desirable social changes—including some kinds of change whose possibility has been cast in doubt by mainstream evolutionary psychology—are achievable; indeed, it suggests that we are a long way from having tested the true scope of human possibility. More practically, this approach has the potential for showing (sometimes, at least) which kinds of change and which methods of change are more or less likely to succeed. There is room for optimism about the prospects for social change, and about the possibility of developing powerful tools for instigating it. But optimists must accept that all changes are not equally possible or easy: evolved human nature does have implications for the possible outcomes and paths of change, and as we learn about what these are, we are likely to find that they raise some difficult moral questions.

The view sketched here also has implications for the conduct of inquiry—and broader public discussion—in areas where evolutionary science, social thought, and political decision making intersect. Claims about the implications of evolutionary psychology for social policy do not usually appear in evolutionary psychologists' research papers but in works aimed at synthesizing a larger body of research and exploring its broader implications. Such works are often written for a nonspecialist audience, sometimes by leading empirical researchers but sometimes by thinkers who become prominent mainly on the basis of their conceptual and synthesizing work. The prominence of these books in discussions of the social lessons to be drawn from evolutionary science has been notable since the publication of Richard Dawkins's *Selfish Gene*, and although some of this "pop" evolutionary psychology has rightly drawn stiff criticism (Kitcher 1985; Panksepp and Panksepp 2000; Dupré 2003; Buller 2005; Richardson 2007), there is also much to praise in the contribution that such works make to a wide public engagement with scientific ideas about human behavior and human society.

The discussion that follows focuses especially on works by E. O. Wilson (Wilson 1975, 1978), Robert Wright (Wright 1994a, 1994b), and Steven Pinker (Pinker 2002), all leading synthesizers of the research in mainstream evolutionary psychology. Their works are widely read, and the view of human possibilities that they present is fairly typical of mainstream evolutionary psychology. But other views can be found in the field itself and in related areas of study. In the latter part of the book, I turn to the work of David Buss, a leading researcher and the author of evolutionary psychology's most widely used specialist textbook. Some of Buss's recent work shows the potential for a fruitful engagement between mainstream evolutionary psychology and broader thinking about the evolution of human sociality and cognition (Buss 2001, 2009; Buss and Schmitt 2011). Such potential is also apparent in Pinker's more recent work (Pinker 2011).

A different and valuable contribution comes from recent discussions about the relationship between feminism and evolutionary psychology (Eagly and Wood 2011). As the philosopher Carla Fehr has recently noted (Fehr 2012), feminist evolutionary psychology is a growing and important area of research; Fehr suggests that deeper engagement with this work by feminist science studies scholars could contribute to better thinking about human evolution and its implications. Some political scientists also see the promise of such engagement: Laurette Liesen (Liesen 2011) calls for feminists to "look beyond evolutionary psychology" to evolutionary biology, developmental biology, and behavioral ecology for insight into human reproductive strategies and their implications. The work of behavioral ecologist Patricia Gowaty on sexual strategies provides a particularly promising instance of evolutionary research informed by feminist insights (Gowaty 2003, 2008, 2011; Moore, Gowaty, and Moore 2003; Gowaty et al. 2007); some of the prospects that it opens up are sketched in the later chapters of this book.

My aim here is, in part, to advance the kind of engagement between social thought and various evolutionary perspectives that Fehr and Liesen recommend. Critics have frequently attacked or dismissed evolutionary psychologists' claims about human nature but have relied too often on purely negative arguments focused on flaws in the evidence offered—a strategy that can lead at best to the negative verdict of "not proven." Where evidence is faulty, of course, it is essential to point this out. But like Fehr, Liesen, Gowaty, and others, I think it is equally vital that feminists and other critics of mainstream evolutionary psychology engage in a positive exploration of what evolution might really have to teach us about human life, human possibilities, and the prospects for change. To bring the enlarged and revised evolutionary psychology envisaged here to full realization and fruitful use will require the establishment of a new kind of discussion, less ideological and more practical. The changes needed to make such a discussion work are explored toward the end of this book; happily, several of them are already under way.

As the next chapter shows, evolutionary psychologists have often written in a way that seems to suggest that we should trim our aspirations for social change to fit the limitations imposed by evolved human nature. This is only one of several areas where critics worry that evolutionary psychologists entangle facts and values by tacitly importing values into their descriptions of human nature and then use the results to support value judgments about how we should behave (Fairchild 1991; Travis 2003). Critics have accused evolutionary psychologists of condoning or justifying many less-than-admirable features of modern societies, including economic inequality, racism, religious bigotry, and rape. Evolutionary psychologists, in turn, have expressed outrage at these accusations, arguing that their aim is simply to get the facts straight so that whatever social choices we make are founded on an accurate picture of what people are like (Buss 1995; Leger, Kamil, and French 2001). Their critics, they say, are the ones guilty of confounding facts and values, rejecting empirical evidence when it conflicts with their ideology, and misinterpreting factual assertions about human nature as value-laden statements of approval or justification.

Evolutionary psychologists thus accuse their critics of assuming that we can reason from claims about facts to lessons about values, from descriptive premises about how things are to normative conclusions about how they ought to be or, indeed, to prescriptive conclusions about what we ought to do. David Hume famously pointed out in the eighteenth century that reasoning that makes this sort of move—deriving an "ought" from an "is," as philosophers like to say—commits a fundamental error. One common way of making this mistake was later dubbed the "naturalistic fallacy." This is the mistake of assuming that because something is natural or occurs in nature, it must be good. When evolutionary psychologists argue that some uncomfortable but common aspects of human societies, such as racism and gender inequality, are natural, critics say, this carries at least an implicit suggestion that these aspects are not so bad after all. Evolutionary psychologists point out, however, that they are alert to the pitfall of the naturalistic fallacy and at pains to avoid it. The facts about human nature, they insist, carry no implications about how things should be. When we identify some human characteristic as part of evolved human nature, this leaves open the question of whether we should celebrate it or strive to overcome it. Many other natural phenomena are harmful to human lives—pests and diseases, storms and wildfires—and scientists rightly seek an accurate understanding of these things to help discover how we can best protect ourselves from them. In just the same way, evolutionary psychologists say, we must face up to the uncomfortable facts of human nature to help us discover how to minimize the suffering they may bring about. And some lessons about the choices we should make do follow directly from the facts—if certain goals are impossible and the resources available are limited, we surely should direct our efforts toward other goals that are more feasible.

But the situation is more complicated than this simple response suggests. First, irrespective of the attitudes that evolutionary psychologists themselves take on such uses, their work is in fact widely cited to support prescriptive conclusions. Whether or not these conclusions should be laid at the door of evolutionary psychology, they—and the descriptive claims that are used to bolster them—warrant examination and response. Second, closer attention to the work of some leading evolutionary psychologists reveals that they themselves do not succeed in keeping judgments about facts and values clearly separate. It should be obvious that descriptive claims can yield substantial normative implications when combined with broad (perhaps tacit) assumptions about values; such reasoning is common in discussions of human nature and the limits of human possibility (Wilson, Dietrich, and Clark 2003). Moreover, certain kinds of beliefs about (merely descriptive) facts are themselves difficult to disentangle conceptually from prescriptive or normative views. Philosophers have noted that some concepts that we use in everyday factual descriptions are inherently value-laden—it might seem that there are facts about whether a particular action was cruel or courageous, for example, but these concepts seem to have a value component that cannot be eliminated, and this sort of entanglement is common in work seeking social lessons from human evolution. Such entanglement between descriptive and normative ideas can have serious consequences, allowing some thinkers to present normative positions as if they are descriptive or derive directly from descriptive facts and are authenticated by evolutionary science. And finally, the idea that many of the goals that would-be reformers have in view can be put aside as infeasible is an idea that can be supported only with the aid of substantial assumptions about values.

Much of my discussion here concerns the tacit value assumptions and value-entangled concepts in play in social applications of evolutionary psychology. There are larger philosophical issues here, though—how to think about values at all from a factual, scientific point of view, and how to think about the place of value judgments within science itself. I return to these questions and to some ramifications of the more practical problem of how to negotiate the interlacing of fact and value in the study of human nature toward the end of the book.

One of the core elements of what evolutionary psychologists see as evolved human nature—by far the most discussed—is sexual differentiation in behavior and cognition (Gowaty 1992; Wright 1994a, 1994b; Mulder 2004; Liesen 2007; Schmitt et al. 2008; Lippa 2009; Fine 2010; Eagly and Wood 2011; McCarthy and Arnold 2011).[1] The idea that female and male brains are evolved to be different (Buss and Schmitt 1993), in conjunction with an adaptive story that explains a long list of putative cognitive and behavioral differences in ways that fit comfortably with widespread gender stereotypes, has been picked up enthusiastically in a burgeoning popular literature. It is used (among many other things) to argue for gender-specific education (on the grounds that girls and boys have naturally different learning styles; Gilbert 2006; Halpern et al. 2011), for reevaluation of the notion of sexual harassment (on the grounds that women and men have naturally different attitudes toward sexual solicitation; Browne 2006), and for rethinking the aims of feminism (on the grounds that women and men have naturally different goals and priorities; Brizendine 2006). More broadly, this idea is used to defend the supposition that some societies are now approaching the natural limits of gender equality—that the gains that women in Western democracies have recently achieved in income, relief from unequal child care obligations, and success in traditionally male-dominated professional and intellectual arenas have gone about as far as nature will allow. This idea has also, of course, been sharply criticized by feminist scholars and others (Ah-King and Nylin 2010). Because this particular area of application is so important and well explored, it serves as a useful focus for this book's investigation. The theoretical framework I develop here is quite general, however, and many of the particular lessons to be drawn from this instance apply equally to other aspects of evolutionary psychologists' conception of evolved human nature.

This book draws together threads from many sources. The debts I owe to earlier thinkers from biology, psychology, philosophy, and economics will be obvious in what follows, as I trace many of these threads and show how they are interwoven. The way that I draw these elements together also has some important precursors. Philip Kitcher's rich body of work at the intersection of evolutionary and ethical thought is one of these (Kitcher 1985, 1996, 2003, 2007, 2011); another is Patrick Bateson's equally wide-ranging work on the interconnections between evolution, development, behavior, and the lived experience of humans and other organisms (Bateson 1983; Bateson and Martin 1999, 2013; Bateson and Gluckman 2011). Working from either side of the boundary between biology and philosophy, Kitcher (a philosopher) and Bateson (a biologist) show how much can be learned by exploring the interplay between questions about human biology and human values.

THE ZOOLOGIST G. G. SIMPSON famously remarked that all attempts to answer questions about human nature and the meaning of human life made before the publication of Darwin's *Origin* are worthless. "We are better off," he concluded, "if we ignore them completely" (Simpson 1969), and a similar attitude is apparent in the work of evolutionary psychologists and its many popularizers. The entanglement of descriptive hypotheses with value assumptions, I argue, suggests reasons to think harder about what evolutionary thinkers might learn from the long pre-Darwinian history of inquiry into these issues by philosophers, political theorists, and social scientists. This study focuses, nonetheless, on the discussion within evolutionary psychology and allied areas of biology; my main aim is to show ways in which the implications of evolutionary science for Simpson's questions have changed since Darwin's time and, indeed, since the time of Simpson himself.

Before considering how evolutionary psychologists think about the implications of evolved human nature for human societies, it will be helpful to get a sense of what their conception of human nature actually is. The remainder of this chapter gives a short overview of some common elements in the various representations of human nature that mainstream evolutionary thinkers have offered over the past few decades. It is not intended to be comprehensive or to take account of all the subtleties that have emerged in this rich literature but to capture the basic conceptual moves that are most relevant to thinking about the limits on human social possibilities. (We turn to some of the subtler points toward the end of the book.) The ideas sketched here have become common currency and should sound very familiar to most readers.

Evolutionary psychologists argue that almost all of the important evolutionary history that shaped human nature took place in an environment quite different from that in which most of us live today. Humans—the genus *Homo*—emerged about 2.4 million years ago, close to the beginning of the Pleistocene epoch. For almost all of its subsequent evolutionary history, our species lived in small mobile bands, making a living by hunting and foraging. The large sedentary social groups in which most people now live did not begin to emerge until after the dawn of agriculture, only a few hundred generations ago. Our patterns of cognition and social behavior thus either evolved or were maintained by natural selection in what evolutionary psychologists call the environment of evolutionary adaptedness (EEA): the social and physical environment typical of Pleistocene hunter-gatherer social groups. This means that some of those patterns are not well suited for life in modern environments. As leading evolutionary psychologists Leda Cosmides and John Tooby famously put it, "our modern skulls house a stone-age mind," and mismatches are inevitable. Our fondness for sweet foods, for example, evolved in a world in which sugars were rare and sweetness marked foods that were valuable for their high caloric density. In today's environment, however, our tendency to prefer sweet foods has become a liability, more destructive than advantageous to our health.

The basic theoretical framework upon which evolutionary psychology builds its picture of the evolution of human nature is that of gene-level selection. A gene can succeed and spread through a population if it has effects that, on average, tend to increase its own representation in the population. The gene can do this directly, by increasing its bearer's individual fitness (i.e., the bearer's capacity to survive and reproduce), or indirectly, by causing its bearer to behave in a way that enhances the fitness of other organisms that are likely to carry the same gene without imposing too large a reduction in fitness on the gene bearer itself. How large a reduction is too large? A gene that causes you to act altruistically toward another person—that is, in a way that increases the other's fitness and decreases your own—will succeed in spreading (on average) as long as the loss of fitness that it imposes on you is less than the increase in fitness it gives to the other multiplied by the probability that the other also carries the gene in question. In sexually reproducing organisms, a rare gene possessed by an organism has about a 50 percent chance of being shared by one of its full siblings or one of its offspring. So a gene that causes an organism to behave altruistically toward a sibling or offspring will spread through the population if, on average, the behavior increases the recipient's fitness at least twice as much as it decreases that of the agent.

All organisms, therefore, must be evolved to be "selfish" in the sense that their genes mostly predispose them to behave in ways that serve the organisms' own fitness—their own ability to gather and protect resources, to attract and keep mates, and to produce and rear young. They may also have genes that predispose them to altruistic behavior, but in most cases these tendencies will be weaker and will fall off rapidly as their degree of relatedness to potential beneficiaries of the behavior decreases. Humans are thus expected to be naturally somewhat self-regarding but willing to make considerable sacrifices for offspring and other close kin as well as lesser sacrifices for kin who are less close. When we have power to determine how goods are distributed, we are likewise expected to show a marked bias in favor of those most likely to share genes with us—our own relatives. Some kinds of altruism or cooperation can also evolve by enabling individuals to work together to increase their fitness in a way that none could achieve alone—hunting large game or protecting an important resource, say—or to reciprocate each other's help, sharing food or child care. We are expected to have a tendency to form alliances that benefit all members and to exchange altruistic acts with others or offer generosity to those who are likely to reciprocate later. Because of the bias toward kin, and because of a weaker bias toward those with whom we have a history of reciprocity, we are expected to have a strong tendency to distinguish sharply between "ingroup" and "outgroup" members, favoring the former and distrusting the latter. It is often thought that this pattern gives rise easily to racism, intercultural antagonism, and xenophobia more generally. When we can gain something that enhances our own fitness or the fitness of members of our ingroup by harming outsiders, even violently, it is expected that we will take the opportunity as long as we can do so safely. Because the threat of retaliation is one of the main reasons to refrain from such behavior, we are expected to have a strong tendency to retaliate against perceived harms and to fear retaliation when we do harm others.

Our ancestors' success depended on their position in the social hierarchy of the group and their ability to negotiate beneficial relationships with other members of the group, especially those whose status was higher than their own. We are thus naturally sensitive to social status and motivated to pursue it; sensitive to social hierarchy and apt to create and strengthen hierarchical relationships. Our ancestors' ability to negotiate social relationships depended on their ability to control the information others had about them—that is, to communicate and to deceive. Because our ancestors stood to gain if they were able to make accurate judgments about others' intentions and trustworthiness, we are reputation trackers and gossips. A reputation for honesty was valuable for our Pleistocene forebears, and easiest to get by actually being honest, yet there are obvious advantages to duplicity as well. We are evolved to be more honest and generous when we are under scrutiny but less so when we feel we can get away with it.

Evolutionary psychologists argue that male and female humans face importantly different selection pressures, and that these pressures have produced sex differences in cognition and behavior analogous to the obvious morphological differences between women and men. The story that evolutionary psychologists tell about these sex differences has been sharply criticized from the time of its first introduction (Kitcher 1985), but although some important adjustments have been accepted by many evolutionary psychologists, the broad outlines of the story remain unchanged and remain at the heart of evolutionary psychologists' thinking about human nature. I offer grounds for a deeper revision of this fascinating but flawed story shortly, but first it is useful to understand it in its standard form.

According to this story, sex differences in cognition, behavior, and morphology all stem ultimately from the basic difference between the sexes as evolutionary theorists understand it: a difference in the size of the gametes. When two organisms mate—combining their sex cells to create the zygotes that will become their offspring—females contribute large, resource-rich ova while males contribute small, mobile sperm, which carry little besides genetic information. This initial asymmetry has driven the evolution, in many lineages, of much greater asymmetries in the "parental investment" of the two sexes. Since females have more to lose if a mating fails to produce successful offspring, they are selected from the beginning to invest yet more resources in order to protect their initial investment. In most species, females contribute more than males do to the rearing of the offspring. In mammals, including humans, part of this investment takes the form of gestation and lactation: mothers support the offspring within their own bodies through the early stages of development and provide milk from their own metabolic stores to nourish infant offspring. This means that mothers are yet more strongly selected to strive to ensure the survival and healthy growth of the children they have borne. Men, who can father children with only the slightest investment of time and resources, face quite a different selective landscape, one in which it is often more productive to invest effort in fathering a new child than in supporting one who is already born.

Some of the sex differences that evolutionary psychologists expect and explain on the basis of the different evolutionary forces acting on men and women are expressed directly in mating strategies—patterns of mate choice and mating behavior. The number of children a woman can bear is strictly limited by the demands of pregnancy, whereas a man can father a much larger number of children if he is successful enough in attracting mates. To maximize the number of children born to them who reach a healthy adulthood, women should seek to ensure that the children they bear have the best possible chance of thriving while men should seek to maximize the number of children that they father, to the degree that they do this without severely compromising the children's chance of survival. This implies that women should choose mates very selectively, preferring those who are likely to help the offspring thrive by contributing either superior genes or extra resources. They are thus expected to prefer older mates with wealth or social status, or perhaps young ones with extraordinary accomplishments signifying superior genetic endowment. Men, on the other hand, should be interested in having plenty of matings with a variety of mates and should prefer mates with the markers of fertility and health—that is, those who are young and physically attractive. Men are thus expected to be more promiscuous than women, more likely to be unfaithful to their partners, and particularly likely to seek new and younger mates when their partners reach the age of declining fertility and their children are past the vulnerable early years.

The sexes are expected to differ as well in their attitudes toward sexual infidelity in their mates. A woman stands to lose most from infidelity if her partner shifts his parental investment to a new family; sexual infidelity alone is a less important threat to her fitness than emotional attachment to a new mate. A man stands to lose most if his partner bears another man's child and he is fooled into investing resources in raising a competitor's offspring. Sexual infidelity alone is a serious threat, irrespective of emotional attachment. Women and men are thus expected to show different patterns of jealousy, with women being especially jealous of their partners' affection and generosity and men being especially jealous of their partners' sexual favors.

The same basic differences in mating strategy are sometimes seen as explaining the "sexual double standard" (the fact that sexual activity is usually far more restricted and socially costly for women than for men) and the "Madonna–whore complex" (the fact that men have very different standards for partners in casual matings than they do for marriage partners). The double standard simply reflects the biological fact that a woman's infidelity is far more damaging to her partner's fitness than vice versa. The

Madonna–whore complex reflects the fact that men are evolved to be very selective in choosing a mate whose offspring they will help support—especially if this requires a limitation of other mating opportunities—but fairly undiscriminating when it comes to those other mating opportunities. Notoriously, the basic differences in mating strategy between the sexes are also seen as explaining the phenomenon and basic asymmetry of rape. For men who would otherwise be unable to find mates, rape is a means of achieving matings. It may be physically and socially risky, but men are expected to be selected for a tendency to rape when the fitness advantages of doing so outweigh these risks, such as when they have no other way of fathering children or when the risks can be minimized, as is often the case for "date rape."

Evolutionary psychologists expect broader behavioral patterns to show differences between the human sexes as well. Women have been selected for behaviors that help ensure that their children thrive, so they are expected to be more nurturing and interested in children than men are. Men have been selected for behaviors that help them win and defend both mates and resources, and so are expected to be more aggressive than women. They have also been selected for behaviors that enhance their social standing or advertise their genetic endowment, and so are expected to be more interested and gifted in public displays of strength, courage, or artistic or musical skill than women are. The stereotypical male "rock star" (literal or figurative) whose displays of virtuosity win him sexual access to crowds of young female admirers is seen as a classic expression of this pattern. For similar reasons, women are expected to tend to focus their interests in the domestic sphere (except when pursuing stellar mates in the public arena) while men are more apt to vie for social and material success in the public sphere. Women can best maximize their children's success by living safely, while men have the chance of vastly enhancing their status and mating possibilities by taking high-stakes risks, so men are expected to be more accepting of risk than women are, even becoming attracted to risk where risk-seeking serves as a rock-star-like display of superior genetic endowment signaling exceptional strength, speed, or nerve.

Some of these examples begin to hint at a further type of difference discussed by evolutionary psychologists—differences in cognitive capacity. It is thought to be likely that human societies in the Pleistocene resembled most modern hunter-gatherer societies in having a well-marked sexual division of labor. According to this picture, women would forage for foods and other resources while caring for children; they would probably do this together for safety and to take advantage of opportunities for cooperation. Men would specialize in hunting, using projectile weapons, and traveling over longer distances in pursuit of game, hunting together for large game but perhaps singly for smaller prey. These different specialized activities would impose different selective forces on the sexes. Women would be selected for social skills, language use, fine motor skills, and the capacity to multitask. Men would be selected for the capacity to create and use tools, to throw projectiles forcefully and accurately at moving targets, and to track game over extensive landscapes without getting lost—that is, for skill in spatial reasoning and in manipulating physical objects—as well as for the special social and cognitive skills involved in planning and carrying out coordinated hunting of large and dangerous animals armed only with sharpened sticks and stones.

The preceding paragraphs offer only the simplest sketch of human nature as seen by mainstream evolutionary psychology; the full picture is much richer, more complex, and more interesting, despite its deep flaws. But this simple sketch is sufficient to capture the main elements in play in most discussions of the limits that evolved human nature imposes on our social prospects. One feature of this description is obvious and important—the repetition of words that indicate probabilities, tendencies, and expectations. How can such an amorphous human nature impose strict limits on the kinds of society we can create? What does it mean to say that it does? These are the first questions that an examination of the social implications of our evolutionary history must consider.

[2]

The Cost of Change

THE IDEA OF LIMITS SET by human nature is sometimes presented in terms of absolute bounds on what is possible for organisms like us. But evolutionary psychologists recognize that some environmental interventions can have large effects and that absolute bounds may therefore be difficult to find. The limits set by human nature are thus often understood as a matter of degree, their practical significance indicated by an economic metaphor: some outcomes, though possible, are said to be achievable only *at a cost*. This notion is sometimes carried further to suggest more explicitly that a cost–benefit analysis would reveal that certain apparently desirable social changes would carry with them costs too high to pay. The use of economic metaphor here poses some difficult conceptual puzzles, and more complete assessment of this way of thinking must wait until the broader evolutionary picture that I advocate has been drawn. But this chapter examines some of the notable appearances of the notion of the cost of social change over the past decades, examining both the functions it serves and the problems it raises in order to set the stage for a fuller appreciation of its importance and limitations. I ultimately argue that the way mainstream evolutionary psychology talks about costs is confused and misleading for several interconnected reasons. This chapter begins to expose two of these reasons: thinkers seeking lessons for human society from evolutionary psychology blur the distinction between facts and values in ways that they do not recognize, and they fail to look clearly enough into the complexity—and the limitations—of cost–benefit analysis. Later chapters examine these failures more closely and show how they are linked to an overly narrow view of the dynamics of evolution. This investigation will move toward a more well-founded conception of the costs, benefits, and trade-offs of social change integrated with a broadened view of evolutionary psychology, but it will also move beyond the cost–benefit approach to note some ways in which that broadened evolutionary perspective indicates the need for other, more nuanced ways of thinking about the comparative value of different ways of living.

A typical expression of the cost of change is given by the journalist and scholar Robert Wright (1994a, 34). Wright combines a claim about absolute bounds and another about costs in a single warning: "People, though malleable, aren't simply and infinitely malleable. They aren't malleable enough to make communism a productive economic system, and they aren't malleable enough to create a society of perfect behavioral symmetry between men and women. Some changes simply can't be made, and others will come only at some cost." Note that with the introduction of the concept of *cost* we have stepped into the realm of the normative—we are not talking just about facts but about values. How to understand this move is one of the main problems that must be confronted in considering the literature on evolutionary psychology. A first step is to clarify the nature of the cost or costs here invoked.

The notion of a "cost of change" imposed by evolved human nature first took clear shape in sociobiology's early days. In a passage that was soon to become notorious, Richard Dawkins warned of the challenge facing any effort to overcome our evolved tendency toward selfishness. But he went on to give a cautiously optimistic view of the prospects for success in meeting that challenge:

> Be warned that if you wish, as I do, to build a society in which individuals cooperate generously and unselfishly towards a common good, you can expect little help from biological nature. Let us try to teach generosity and altruism, because we are born selfish. Let us understand what our own selfish genes are up to, because we may then at least have the chance to upset their designs, something that no other species has ever aspired to. . . . Our genes may instruct us to be selfish, but we are not necessarily compelled to obey them all our lives. It may just be more difficult to learn altruism than it would be if we were genetically programmed to be altruistic. (Dawkins 1976, 3)

Dawkins thought change might be possible, with effort. But the other great spokesman for early sociobiology, E. O. Wilson, saw effort as only part of the picture. He distinguished two kinds of costs that might attend social change of the "more difficult" sort that moves us away from our natural patterns of behavior—the investment of effort required to push human behavior beyond its natural limits and a further cost in human well-being:

> Human nature is stubborn, and cannot be forced without a cost. . . . We now believe that cultures can be rationally designed. We can teach and reward and coerce. But in so doing we must also consider the price of each culture, measured in the time and energy required for training and reinforcement and in the less tangible currency of human happiness that must be spent to circumvent our innate predispositions. (Wilson 1978, 147–48)

Under the heading of "time and energy required for training and reinforcement"—the investment of effort—we might further distinguish two components: the one-time investment required to make the initial transition to new social arrangements and any ongoing investment needed to maintain them. Wilson does not separate these, but for the sort of cost–benefit analysis he seems to be advocating, the distinction is important.

Wilson's further idea that the "less tangible" costs to human well-being are crucial has been picked up more recently by the cognitive psychologist Steven Pinker. Pinker connects concerns like Wilson's with more concrete political worries:

> The issue is not whether we can change human behavior, but at what cost. . . . We cannot pretend that we can reshape behavior without infringing in some way on other people's freedom and happiness. . . . Inborn human desires are a nuisance to those with totalitarian and utopian visions, which often amount to the same thing. What stands in the way of most utopias is not pestilence and drought but human behavior. So utopians have to think of ways to control behavior, and when propaganda doesn't do the trick, more emphatic techniques are tried. (Pinker 2002, 169–70)

Michael Ruse's recent overview of the philosophy of human evolution (Ruse 2012) echoes (somewhat cautiously) the ideas of Wilson, Pinker, and Wright: "What the Darwinian is likely to say is that one should be cautious about utopian proposals for complete sexual identity. . . . It might just be that women want to spend time with their young children in ways that men do not. The Darwinian might say that the moral course of action is not to pretend that this is not the case, or to try to brainwash people out of it, but to reorganize society so that these natural desires can be fulfilled for the best ends of all" (196).

The concern these authors express about the cost of social change does not imply that they are opposed to attempts to eliminate discriminatory cultural practices or institutional structures where these prevent people from exercising their natural preferences and realizing their natural capacities. On the contrary, most of them explicitly note their support for this sort of social change and the equality of opportunity that it aims to bring about. But where the change sought is one that they see as opposed by human nature—such as creating a society with "equality of outcomes" between the sexes, one in which half the mathematicians, scientists, legislators, and chief executive officers (for example) are female, and half the child care is done by males—they hold that serious efforts to realize this change are likely to involve costs so high as to be morally unacceptable.

Pinker implies that the only way to override people's inborn desires is by means of the "more emphatic techniques" characteristic of totalitarianism. "No one doubts that people's behavior can be controlled," he notes; "putting a gun to someone's head or threatening him with torture are time-honored techniques" (Pinker 2002). He concludes that the cost of such interventions is measured partly in unhappiness and partly in the loss of freedom. Ruse seems to take a similar view, suggesting that natural desires could be overcome only by "brainwashing," while Wright's insistence that humans are "not infinitely malleable" appears intended to answer the contrary claim—made by the totalitarian torturer in George Orwell's *1984*—that "men are infinitely malleable" and therefore amenable to whatever terrible reshaping a "utopian" could desire. Pinker hints that many of the ideals he regards as utopian are ultimately entirely unrealizable (hence the label), but he thinks that a serious effort—however futile—to create a society that breaks away from the pattern set by our innate desires and predispositions must result in a significant abridgment of freedom. The implication is not just that some outcomes that feminists and other "utopians" have hoped for are beyond human reach but that it is a dangerous mistake even to pursue them.

In the words of the conservative political philosopher Michael Oakeshott, approvingly quoted by Pinker: "To try to do something which is inherently impossible is always a corrupting enterprise" (2002, 290).

Pinker goes on to suggest that the dangers of pursuing impossible dreams extend further, threatening not just liberty and happiness but also the realization of human capacities. His idea here is familiar from dystopian fiction (Pinker points to Kurt Vonnegut's classic story "Harrison Bergeron" [Vonnegut 1961]): that there are trade-offs among the desirable characteristics that can be realized in human populations, so that the special environments needed to achieve some such characteristics inevitably impede the realization of others. Greater equality in outcomes on some measure, for example, might be achievable only at the cost of lowering the peaks on the same measure: "Policies that insist that people be identical in their outcomes must impose costs on humans who, like all living things, vary in their biological endowment. Since talents by definition are rare and can be fully realized only in rare circumstances, it is easier to achieve forced equality by lowering the top (and thereby depriving everyone of the fruits of people's talents) than by raising the bottom" (Pinker 2002, 425).

In Vonnegut's story, exactly this effect is produced—by a government bent on producing perfect equality—by crudely handicapping the gifted, strong, and beautiful so that they are effectively no better than anyone else. For example, in the real world (Pinker seems to mean), if most of the children who have extraordinary mathematical talent are boys, we would probably undermine their chances of fully realizing those talents if we comprehensively reformed the teaching of mathematics in ways that make it more effective for (less talented) girls (Kane and Mertz 2012). Another sort of trade-off might arise where better performance on one measure is achievable only at the cost of diminished performance on a different measure, perhaps (but not necessarily) for the same individuals. It is possible, for example, that the environments that foster higher levels of mathematical or athletic achievement in some women might hinder the fullest expression of their capacity for compassion or linguistic performance—or perhaps that of other women or of men. These are (possible) trade-offs in the range and distribution of human phenotypes—i.e., in the characteristics that humans actually come to express as a result of the influence of both genes and environment—so I will term them *phenotypic trade-offs*.

So the putative costs of trying to modify human behavior include (1) the cost of investment in transition to and maintenance of a new set of social arrangements, (2) the cost of reduced happiness, (3) the cost of reduced freedom, and (4) the cost of phenotypic trade-offs at either the individual or the social level.[1] These categories are barely indicated in the brief explicit discussions of cost that evolutionary psychologists offer, however, and any serious attempt to assess them would need to undertake a complex accounting of many aspects of cost that go unmentioned. Many of these aspects are closely connected with assumptions about the causal processes that shape behavior, development, and evolution, so a fuller treatment must wait until the relevant assumptions have been examined.

Some of the issues that need attention can already be identified, however. Cost–benefit analysis is an economic tool for comparing possible plans or policies, and for comparing them either to the status quo or to each other. Cost–benefit analysis is rooted in the utilitarian idea that all goods and harms can be reduced to some sort of comparable units of happiness or "utility," and that the net number of these units gained is the only measure of value that can be used to judge between alternative possible courses of action. In classical cost–benefit analysis, monetary units such as dollars are used to stand in for the underlying utility units (what philosophers sometimes call "utiles," or "hedons" to acknowledge their connection to hedonic pleasure). But in any project that involves practical human life at all, there are bound to be good and bad effects that are not readily measured in monetary terms—standard examples of "nonmonetary goods" include longer life expectancies, social cohesion, fertility, aesthetic value, health, and autonomy. A similar problem affects cost–benefit analysis understood in the looser sense relevant here, in which (presumably) we are concerned not with dollar values but with the broader positive and negative effects of social choices. How are these (doubtless quite diverse) effects to be evaluated and compared?

Key questions for any serious attempt to use cost–benefit analysis to assess proposals for achieving social change include the following:

What factors ought to be included in the cost–benefit calculus, and how ought they to be weighed? The quotations collected above use the word *cost* but don't talk directly about corresponding benefits. Cost–benefit analysis obviously requires that benefits be included in the assessment, not costs alone. What can be said about the benefits of the kinds of change that evolutionary psychology warns about? Both costs and benefits will be unevenly distributed across groups within the larger population. How ought we to weigh costs and benefits that affect different people differently? Under conditions of environmental transformation (such as climate change), the failure to realize appropriate social change can itself have high costs. This means that the concept of the "*status quo ante*"—leaving things "as they were before"—becomes problematic. How can these costs of adaptive change foregone be included in the calculus?

What aspects of how the change develops over time need to be considered, and how? As mentioned earlier, the one-time investment required to make a transition must be separated from any ongoing costs of maintaining the new state. But this is not enough. Transition itself is not instantaneous and may be a long process. What structural features of the transition process need to be considered, and how? For example, is it a gradual process with increasing effort required to push through the final stages? Is it a struggle in which things get much worse before they (possibly) get better? Is there a "tipping point" so that change becomes self-reinforcing and swift after a certain point? Or does the transition have a structure different from any of these common patterns? It is normal, in cost–benefit analysis, to "discount" future costs and benefits to some degree on the principle that a bird in the hand is worth two (or at least, more than one) in the bush, so the speed at which change comes to fruition is important. The discount rate for delayed costs and benefits also needs to be determined. How much less are benefits (or harms) worth if they will not arrive for another five years, or another five generations?

What kinds of interventions are involved in bringing change about, and what are their cost implications? Costs depend on the nature of the intervention, so we need a clearer idea of the kinds of interventions envisaged. Are they applied to internal or external factors—that is, do they change people's behavior by modifying their "inner" states (their beliefs, desires, goals, fears) or by modifying the contexts in which people find themselves? (This isn't a sharp distinction, but it matters. A simple example is the behavior of people waiting for a bus. Do they form a first-come-first-aboard line, or push and shove at the last minute? You can try to get people to line up nicely by teaching them about fairness or threatening them with fines, or you can do it by installing a post-and-cord structure that cues and scaffolds the formation of a line. This last approach—an external one—is much easier in this case. But sometimes changing people's minds is the most effective way to change what they do.) Complex systems such as organisms and societies often have leverage points where a small effort can have a large effect. (The post-and-cord crowd-control device just mentioned is a simple example of an intervention at a leverage point, but changing "hearts and minds" can be another sort of example. We will see some other examples later on.) Are there levers that could increase the amount and the speed of social change? What would that do to the cost–benefit balance?

Are there special problems with applying the economic concepts of cost *and* benefit *in the much broader context of large-scale social change?* The notion of investment or expenditure is not easy to apply in contexts of broad social change, where the investment in question consists of human activity that may have intrinsic positive and negative values to the agents. For example, some of the costs mentioned by Wilson take the form of "effort"—that is, of people doing things. But what if the people who make the effort enjoy the activities involved or value the opportunity to contribute? The usual cost–benefit fiction that all costs and benefits can be compared by counting them in monetary terms makes even less sense here than it usually does. How can the costs of all the different kinds of human activity involved in any social change be evaluated? Who is to assess the costs and benefits associated with the process and outcome of change, and by what means? The danger of enlisting experts to decide what other people really want is obvious. It seems essential to canvass people's own views about how they value the activity and effort they are called upon to engage in. Yet such self-valuation faces serious difficulties of its own, well-attested in the literature on cost–benefit analysis. People are apt to overlook certain kinds of goods and harms in decision making about their own lives and may overemphasize others. A final problem is perhaps most difficult of all: How ought we to calculate costs and benefits when the changes under consideration may themselves change people's assessment of the value of different outcomes by changing the thinking of the people themselves?

The concept of human nature might be thought to resolve the problems of assessment noted earlier by showing how some outcomes are objectively better for people than others, whether the people involved recognize it themselves or not. Some proponents of evolutionary psychology as a guide to social policy certainly seem to believe something like this. But the idea of human nature is itself problematic, owing to the hazy boundary between descriptive and normative conceptions of human nature and human variation. Some of the haziness here is ancient. Aristotle saw any thing's nature as determining its good, so that to bring true human nature to full realization was the highest good to which humans could aspire, whereas in the Christian tradition human nature is fallen and sinful and must be overcome to reach the divine good. Rousseau and the Romantics saw human nature as innocent, free, and loving while Hobbes and his followers in the "dismal science" of economics saw it as fundamentally self-interested. Scientific conceptions of human nature have sought a value-free understanding of our thought and behavior, but their efforts have shown a persistent tendency to slide over into seeing human nature as either good or bad, pure or corrupt. Value judgments are important (of course) to good decision making, but when they are unexamined and unrecognized, hidden under the guise of objective science, they can be seriously misleading. The conceptual territory here is therefore fraught with hazard.

The way that many evolutionary thinkers talk about the costs of change already reveals a blurring of the boundary between descriptive and normative conceptions of phenotypic patterns. Some phenotypes (including behaviors, capacities, or desires) are accepted as "natural"; others are seen as "forced." This distinction is reminiscent of Aristotle's separation between the "natural" motion of objects (falling, for heavy bodies; rising, for flames) and the "violent" motion brought about by external forces (like the upward motion of a cannon ball). In keeping with this Aristotelian spirit, "forced" phenotypes are expected to result in unhappiness for the organism involved. "Human nature is stubborn, and cannot be forced without a cost," said Wilson (1978, 147); the kind of force required is indicated by Pinker ("more emphatic techniques" [2002, 170]) and Ruse (brainwashing).

A related contrast appears in the broader literature on evolution and sex differences in the context of explanations of particular phenotypic outcomes. Some phenotypes are explained simply as expressions of evolved human nature; others require special explanation by appeal to unusual environmental factors. Stereotypically "male" preferences or capacities in some girls and women, for example, are explained as resulting from the "masculinization" of female fetuses by abnormally high levels of testosterone in the uterine environment. The point here is not that these explanations are mistaken in pointing to an important role for prenatal hormone exposure but that such hormones also play a role in "typical" development processes that is as important as their role in "atypical" ones; appealing to them only to explain deviations from the "natural" course of development suggests, again, a kind of Aristotelian essentialism about developmental patterns (Dupré 1998). Sometimes this essentialist approach is carried further so that extreme types, rather than average ones, are considered typical or natural, as when "male brains" are said to be "hardwired for competition" and "female brains" "hardwired for empathy" (despite the manifest fact that both sexes display both capacities), or when autism is described as symptomatic of "the extreme male brain" (Baron-Cohen 2002). Here relatively small (putative) differences in average phenotypes between the sexes are treated as imperfect manifestations of male and female "natures," which are seen not as modestly different in certain respects but as polar opposites.

A final and important problem for any assessment of the cost of change is to determine the scope of change that is actually under consideration. In warning against reformers' efforts to overcome "human nature," evolutionary psychologists focus their criticism on goals so ambitious that Pinker's "utopian" label seems easy to justify: goals such as the creation of a society in which everyone cares for strangers exactly as much as for close kin, or one in which the behavior of men and women is indistinguishable except where their bodily differences force them to behave differently. Pinker has a special worry about utopian goals, based on his reading of the cost–benefit calculus that they entail. If you are a utopian in the sense of believing that a perfect world—a literal Heaven, or a heaven on earth—can be achieved by human effort, you should be willing to make extreme sacrifices of your own and others' well-being to achieve it. If the end justifies the means, as cost–benefit thinkers take for granted, a superlatively good end justifies the most terrible means. Pinker thinks that this kind of utilitarian reasoning distorted by utopianism is responsible for many of the most hideous atrocities of human history, from the Spanish Inquisition to the Soviet gulags. It is undoubtedly true that utopian goals in this sense have sometimes been used to justify morally heinous acts. I am much more doubtful than Pinker is about what causes the bad moral choices in these cases. Are utopian beliefs the cause, as Pinker suggests? Or are they merely rationalizations for actions that have quite different motivational roots in responsive human nature? It is noteworthy that many of those willing to harm others in pursuit of "utopian" goals have been unwilling to make radical sacrifices themselves.

The point of practical importance, here, is more mundane: that while extreme or even "utopian" goals have sometimes been pursued, it should be obvious that they are not what most social reformers have in mind. Those who advocate measures aimed at achieving human possibilities that are yet unrealized—minimizing the effects of racism and sexism, increasing our capacity to feel compassion for people in distant places and to solve problems in just and peaceful ways—are not in general utopians in any strong sense (though of course some are); they are optimists. The question is what degree of optimism is justified, and on what grounds. Advocates of evolutionary psychology often have an ambivalent attitude about the prospects for social change, hoping for social progress yet doubting that deep change in human social arrangements is feasible. They begin by criticizing utopian goals, but the lessons they ultimately draw are often much broader, suggesting that attempts to bring about even modest further social change may be futile and even dangerous. Evolutionary psychologists' own positive hopes are noteworthy, and I explore them further in chapter 8. It is their pessimistic conclusions that need most careful scrutiny.

[3]
Thinking About Change and Stability in Living Systems

IN THE DECADES SINCE SOCIOBIOLOGISTS first argued that human nature stands in the way of some human aspirations, their critics have repeatedly emphasized an important and obvious point: that we cannot really know the limits of human possibility without knowing the relevant response functions (or what biologists sometimes call "norms of reaction") in their entirety—that is, without knowing how *every possible* environment will affect the expression of each trait of interest—and we do not know this now and never shall. But though the point is correct if taken literally, we can nonetheless have very good grounds for concluding that certain phenotypes really are unachievable for organisms with the genetic makeup of today's humans, and that other phenotypes require environments that are difficult to create or maintain or that have unacceptable effects on other traits. To take a mundane example, we have reasonable grounds for believing that a world in which everyone lives to the age of five hundred is impossible to achieve without genetic modification or quite radical environmental manipulation, and that a world in which almost all people live to the age of one hundred might be achievable but would require an extraordinary investment of resources and involve obvious trade-offs, including some that would result in losses of freedom and happiness. We can make these claims with confidence because of what we know about the human response functions relevant to longevity and their relationships to some others. We know that maximum longevity is fairly consistent across a wide range of favorable environments; that it diminishes in environments with certain kinds of nutritional deficits or excesses, infectious agents, and environmental toxins; and that it increases in environments of certain kinds but only within quite predictable bounds. (For most of these patterns we know a good deal about the underlying mechanisms that make them so reliable.) We also know that there are trade-offs between longevity and certain other phenotypic traits and behaviors—freedom to engage in hazardous sports, to take the performance-enhancing steroids that many athletes use, or to consume alcohol or unhealthy foods, for example, would be curtailed in a society that assigned the highest priority to maximizing longevity. Such a society would thus have lower peaks than ours does on some measures of athletic performance and perhaps of creative output, and its parties would be duller. (New prospects for developing medical techniques for intervening directly in the mechanisms that control aging do open up other possibilities here. We'll recur to this idea later on, but for now note that even these possibilities seem likely to involve considerable investment and significant trade-offs, should we attempt to realize them at a global scale.)

These assumptions about the human response functions relevant to longevity seem unexceptionable. Mainstream evolutionary psychologists think that they can make similarly well-supported assumptions about the human response functions relevant to sex-differentiated cognitive capacities and behaviors, and about many other human cognitive and behavioral features as well. Understanding the assumptions these thinkers make is thus crucial to assessing the claim that we are approaching at least some of the limits of reasonable human aspiration.

Internal and External Causes

To ask what is possible for humans, and at what cost, is in part to ask about the shape of human response functions: that is, about how internal (genetic) and external (environmental) causes would combine to produce the properties of the organism, including its behavior, across the range of possible environments in which it could develop. The response function for each genotype maps its possible environments onto the phenotypes that would result. Our suppositions about the shapes of human response functions (i.e., the range of human possibilities) and about the interrelations among the response functions of different human traits (i.e., the trade-offs among different desirable human possibilities) thus depend on how we think about genetic and environmental causes and how they combine.

The problem of how to think about internal and external factors in explaining how things change or remain stable is both ancient and general. The philosopher Peter Godfrey-Smith sketches three basic ways of explaining change or stability in an object's properties:

Internalist explanations appeal to causes originating inside the object.
Externalist explanations appeal to causes originating outside the object.
Interactionist explanations appeal to a combination of internal and external causes.

What counts as internal (or as external) obviously depends on the object under consideration, so a theorist who takes an internalist view in one context may (indeed, must) take an externalist view in others. In evolutionary explanations, the idea that something internal to biological lineages fully determines their evolutionary path would be pure internalism. Lamarck's pre-Darwinian theory that organisms have a sort of inherent drive toward greater perfection that expresses itself in evolutionary change would be close. The view that external forces such as natural selection do all the work, unimpeded by internal constraints, would be pure externalism. Most evolutionists today are closer to the externalist end of the spectrum, though internal factors have been getting more attention lately in the work of evolutionary developmental biologists and (especially) "process structuralists," who argue that the structure of an organism itself creates the conditions for particular kinds of evolutionary change (Buss 1987; Goodwin and Saunders 1992; Webster and Goodwin 1996; A. Wagner 2014; G. P. Wagner 2014).

In explanations of human development and behavior, the situation seems less settled. Absolute genetic determinism would be a purely internalist position and absolute environmental determinism a purely externalist one. In fact, of course, everybody agrees that both genes and environment influence phenotypic and behavioral outcomes, but theorists nonetheless disagree substantially about the roles played by these components. If we are all interactionists, then interactionism seems to encompass a spectrum of positions ranging across a good deal of the broad span between the poles of pure internalism and pure externalism.

In fact, even the image of a spectrum here is a serious oversimplification. It would be very difficult to defend as uniquely correct any particular way of arranging the various possible combinations of genetic and environmental influence that lie between the poles in a simple one-dimensional order. It is possible, though, to discern some broad categories among them that should be considered more closely.

The simplest way for two factors to combine is additively, so that the same environmental difference has the same effect on different genotypes and in combination with various other environmental differences, and the same genetic difference plays out in the same way in different environments and in combination with various other genetic differences. The magnitude of each effect is therefore proportional to the magnitude of the cause.[1] (For example, a large phenotypic difference between two organisms indicates a large genetic or environmental difference, or a combination of lesser differences in both.[2]) To take a very simple example, consider a plant species that has two genotypes, G1 and G2, and lives in two different environments, E1 and E2. The plants vary in height. For each genotype, plants will always grow taller in E1 than in E2. In each environment, a G1 plant will always grow taller than a G2 plant. A G1 plant in E1 will be the tallest of all, and a G2 plant in E2 the shortest.

Additive combinations of causal factors are easy to understand and model, but genetic and environmental factors do not in general combine additively. In our simple example, for instance, it is perfectly likely that the each genotype would thrive best in a different environment, so that G1 produced taller plants in E1, say, and G2 in E2. Addition does provide an important benchmark for understanding more complex combinations, yet it is difficult even to conceptualize the vast array of possible nonadditive combinations that result when we consider genotypes, environments, and phenotypes with far more than two possible states each. To understand these possibilities, we need some constraints to tame their multiplicity. Views about the interaction of biological causes are often expressed in terms of simple conceptual models or more concrete metaphors that focus attention on important subsets among the possible combinations. The well-known metaphors of "malleability" or the "blank slate" are used to represent a kind of interaction close to the externalist extreme in which the genotype offers little resistance to the influence of the environment (Pinker 2002). The metaphors of "programming" or "hardwiring" represent one close to the internalist extreme in which the genotype fixes an internal "nature" that is resistant to modification by outside forces.

Some thinkers have argued that we need to abandon all such metaphors in order to think clearly about development and evolution (West-Eberhard 2003). I am not sure that this is feasible, and it is clear in any case that these metaphors continue to shape the thinking of both evolutionary psychologists and their critics, affecting the kinds of empirical evidence they seek and how they interpret it. Indeed, a stronger claim might be made here. Cognitive psychologists have shown that our thinking about abstract relations among ideas, and about complex causal relations among objects, are often organized by simple metaphors based on our experience of living in the kind of bodies we have, in the kind of world we live in (Lakoff and Johnson, 1980). These metaphors and the larger conceptual "frames" that they bring along with them are essential but tricky cognitive tools. They help us to reason

FIGURE 3.1 Response functions or norms of reaction are given partial representation by simple diagrams that show for each genotype a curve corresponding to the phenotypes it produces across the range of its environments. These diagrams are very selective: they plot phenotypic outcomes for one or more genotypes on a single phenotypic measure (e.g., height, number of leaves, age at first mating, mathematical skill) against a single environmental variable (e.g., average temperature, nitrogen levels, mother's educational level, prenatal testosterone exposure). They normally show only the actual range of environments, not the full range of possible-but-unrealized ones.

A. Genetic determinism or internal causation—Each genotype (G1 and G2) yields the same phenotypic outcome irrespective of the environment.
B. Environmental determinism or external causation—Phenotypes vary with environment, but in any given environment both genotypes produce identical phenotypic outcomes. (Note that the curve here could be any shape—what matters is that the G1 and G2 curves are the same. The sort of function shown—linear, with a nonzero slope—corresponds to the sort of relation between environmental variation and phenotype that people most often have in mind when they talk about environmental determinism, however.)
C. Addition—Phenotypes vary with both genotype and environment. Response functions for both genotypes are linear, and their slopes are the same: the same environmental change has the same effect on phenotype irrespective of the genotype or current environment.

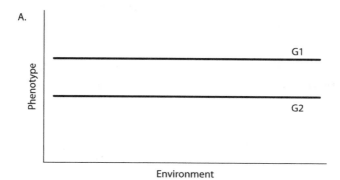

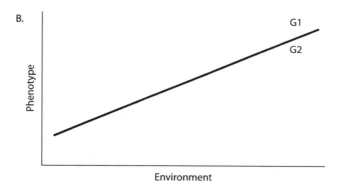

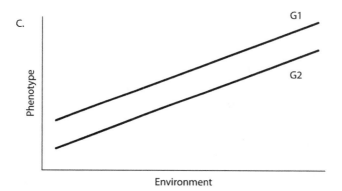

about abstract, complex, and unfamiliar things, but they also carry systematic biases. In particular, they can invoke emotional responses, importing a sense of value into our thinking in a way that may be very difficult to for us to override. Certainly, as the geneticist Richard Lewontin noted long ago, "the price of metaphor is eternal vigilance" (Lewontin 2002, 4).[3] We need, at the least, to scrutinize these metaphors carefully, to consider their limitations and the unintended implications they may carry, to abandon any that are systematically misleading, and perhaps to find new ones that help reveal neglected features of the causal processes of behavior, development, and evolution.

Begin with the basic conceptual models associated with external and internal causes. Externalist explanations employ two basic models of change, one for change in individual objects and one for change in populations of objects. The first is direct action or direct impression. Here some element in the environment directly modifies some property of the object—as impact with another ball changes the velocity of a billiard ball or as pressure from a signet changes the shape of a wax seal. In biological development, the instances that best fit the metaphor of direct impression are cases like the shaping of a tree by constant wind pressure or deliberate training by a gardener. For humans, instances might include injury, body modifications such as piercing or foot binding, or "instructional learning" such as classical conditioning. Importantly, the metaphor of direct impression suggests that any modification is equally easy to achieve, just as the wax seal can take any impression equally readily. The second metaphor for external influence is that of sorting or sifting. Here some structure in the environment systematically removes (or preserves) members of a collection that have certain properties, thus changing the composition of the population. All selection processes fit this basic model, including evolutionary selection but also importantly including selective processes in individual development, such as neuronal selection in brain development, or trial-and-error learning. Here again the metaphor suggests that any selection is equally easy to achieve.

Internalist explanations employ a core conceptual model that takes a variety of metaphorical forms: that of the step-by-step unfolding of some preexisting structure or following-out of a preexisting plan, as in the literal unfolding of a newly emerged butterfly's wing or the execution of a "program" or function for which the organism is "hardwired." Here only one outcome is possible—the full expression of what is already there, laid down and ready to go—unless some outside influence diverts or obstructs the process of unfolding or execution. As the language of "diversion" and "obstruction" indicates, the unique preferred outcome can easily seem to take on a normative status as the correct fulfillment of the object's potential.

These basic conceptual models are adjusted or modified to produce some of the most common ways of thinking about how internal and external factors interact. Wright's claim that "human nature is malleable, but not infinitely malleable" suggests that the internal causes ensure that human development tends toward a particular outcome—an outcome that can be modified by external forces but only up to a point. Variants are possible, but not all are equally easy to achieve, and some are not achievable at all; human nature appears not as wax or clay but perhaps as a rather elastic putty with a resistant core. A related view sees development as internalist unfolding that is subject to significant constraints or interference from the environment. The butterfly's unfolding wings may be bent or torn, the program may suffer from some minor glitches in execution, but the basic pattern remains. A version made popular by Richard Dawkins is the metaphor of the recipe: the genes provide the recipe, the environment provides ingredients and equipment; and the phenotype emerges from their interaction like a finished cake.

These various metaphors of interaction are importantly different from one another, but they share a strong normative flavor: the role of the external factors is either to enable the internal causes to unfold as they ought or to prevent them from doing so. A recipe carries information about how the cake is supposed to come out and about what the ingredients, pan, and oven should be like. If the cake falls, failing to come out as it should, the oven was the wrong temperature. The metaphor of genome-as-recipe implies that the genome similarly picks out a preferred phenotype and thus also picks out the preferred range of environmental specifications that will produce that phenotype. The conceptions of interaction expressed by these intermediate metaphors thus inherit from internalism a normative attitude to both phenotype and environment. From externalism they inherit a tacit assumption that the same external causes will have similar effects on outcomes even when the internal causes are different—too hot an oven will scorch any cake (this assumption is further encouraged by the internalist idea that the role of external causes is merely to enable or obstruct the action of internal ones). Internally caused differences will therefore tend to carry over from one environment to another—if one recipe produces a sweeter cake than another, that difference will tend to remain despite variations in cooking conditions and ingredient quality, for those variations affect both cakes more or less alike.

The common core of these widely used metaphors of interaction can be captured in a general description of what I will call *conservative interactionism*. Conservative interactionism explains an object's features or behavior by referring to causes originating both inside and outside the object. It assumes that internal causes determine a "preferred" outcome and external causes modify it, but not additively: the internal causes tend to keep the outcome close to the preferred state, and large external forces are required to produce significant divergence from it. Extremely divergent outcomes cannot be reached no matter how powerful the external causes. The same external causes modify outcomes similarly for different internal causes. Here the interaction behaves more like pure internal causation than like an additive combination—this is the sense in which the interaction is conservative. A picture like this is clearly consonant with the kind of thinking about the limits and costs of change that evolutionary psychologists have made popular, with their treatment of certain phenotypes as "natural" or "typical," and of divergence from these as deviation requiring special explanation. But Lewontin (again, long ago) has pointed out that a different kind of interaction is possible and may be common (Lewontin 1974). I call the explanatory approach that he articulated *radical interactionism*. Radical interactionism explains an object's features or behavior by referring to causes originating both inside and outside the object. It assumes that nonadditive interaction between internal and external causes may be both strong and multistranded so that external causes could have radically different effects (and effects of dramatically different magnitudes) depending on the internal causes with which they were interacting.

Such a view is not ruled out by anything we have seen so far. But we also have not seen any strong reason to adopt it, and the abstract description just offered is incomplete. Why should we expect organisms to work like this? What kind of patterns might we expect to find among such interactions?

Some important instances of radical interaction can be understood through a conceptual model that has been available to us all along in the form of another metaphor: *response*. Where the other metaphors cast organisms as passively impressionable lumps of wax or rigid mechanisms, the metaphor of response sees organisms as agents capable of sensing the state of particular external variables and actively modifying themselves in accordance with what they sense. The active self-shaping of a houseplant in

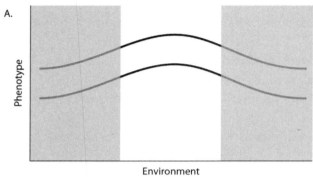

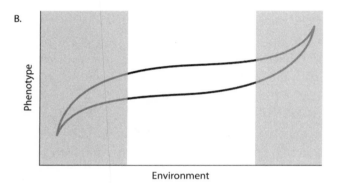

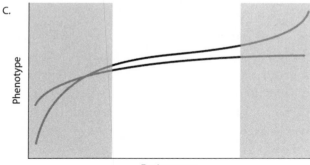

a window "reaching" for the light is a simple example of this sort of interaction. Many kinds of learning seem to have this structure, in which the organism is especially sensitive to (and may actively seek) stimuli of a certain sort and uses them to fix certain aspects of its own behavior or cognitive state. Mammals easily learn to associate particular scents or sounds with feeding and to associate particular flavors with nausea and use these associations to guide their behavior. Young children are acutely sensitive to the vocalizations of the older humans around them and learn to imitate the words they hear. Where these sorts of interactions are in play, multiple outcomes—possibly quite different from each other—are possible; which one is realized may depend on subtle environmental cues. Plants in different circumstances grow differently to maximize their access to sunlight; children in different communities learn different languages. There seems little room for normativity here; it is difficult to see on what basis one outcome (and its triggering environment) could be picked

FIGURE 3.2 Three pairs of response functions that fit the pattern of conservative interaction. The shaded areas indicate environmental states that are rarely or never observed. The phenotypes for both genotypes produce curves that are fairly flat through the range of common ("normal") environments, approximating the internal causation of figure 3.1A.

A. Both genotypes give maximum phenotype values toward the middle of the common environmental range and fall off together in more extreme environments. Changing the environment cannot produce phenotypes that are higher or more equal than they are in currently common environments.
B. The genotypes give quite different phenotypic outcomes across the range of common environments but converge in more extreme environments. Phenotypes far above and below the "normal" range are possible, but only in environments that are not often encountered. Dramatic change to phenotypes and greater equality between the outcomes for the two genotypes can both be achieved, but only by imposing large environmental changes to extreme values.
C. Phenotypes for the two genotypes are close together through the range of common environments but diverge sharply in extreme environments, crossing at one end. Large changes in phenotype can be achieved with large environmental modification, but only for one genotype, so maximizing the peak outcome (or minimizing the lowest) means increasing inequality between the genotypes.

A.

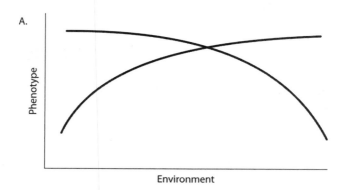

B.

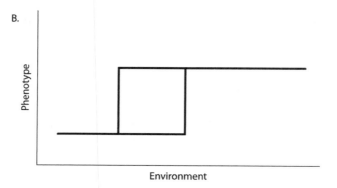

C.

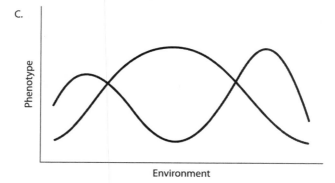

out as the "right" one unless on the basis of a measure of fitness, and different outcomes may be equally fit. Diversity of outcomes thus appears as the result of diverse but (often) equally good responses to diverse environments.

Evolutionary psychologists often use a different metaphor for talking about interactions of this sort, describing them as the playing out of programs that have a complex branching structure whose various conditional routines have environmental triggers. This way of thinking interprets response as a sort of complex unfolding; it lacks the imputation of agency that the metaphor of response carries but does greatly weaken the normative flavor of the metaphors of conservative interaction.

Some theorists, such as Susan Oyama and Mary Jane West-Eberhard, travel further down the path Lewontin indicated, going beyond radical interactionism to reject the basic metaphor of interaction entirely (Oyama 1985, 2000; Oyama, Griffiths and Gray 2001; West-Eberhard 2003). They argue that this metaphor

FIGURE 3.3 Three pairs of response functions that fit the pattern of radical interaction. The examples suggest the range of very different forms that radical interaction can take. What they have in common is that the effect of a given environmental change depends on both the genotype and the existing environment.

A. Two genotypes respond consistently, but reciprocally, to change in the same environmental variable.

Example: One ivy genotype grows larger leaves in shadier environments, the other in sunnier environments.

B. Two genotypes switch between two robust pathways in response to changes in the same variable but at different thresholds.

Example: Two alligator genotypes switch between male and female phenotypes at different temperatures.

C. Two genotypes have different nonlinear responses to the same environmental variable.

Example: One salmon genotype has a higher survival rate of fry at moderately high or low water temperatures; the other maximizes its fry survival rate at medium temperatures, but both have lower survival rates in extremely warm or cold water.

mistakenly pictures the process of development as an ongoing confrontation between two interactants (the genotype and the environment) that retain their identities unchanged throughout. These theorists advocate, instead, a view of development as true ontogeny or coming into being: a sequence of responses, indeed, but one in which what responds at each stage—a functioning genome or a living organism with particular properties—is itself a product of the last stage's response.

The question is not which of these conceptual models captures "the truth" about the roles that genes and environments play in development and behavior. If patterns of change that fit the metaphors of impression, sorting, unfolding, response, or ontogeny occur among the processes giving rise to full-grown organisms and fully realized behaviors—and I think they all do—they all work by means of simple physical causation at the micro level. Moreover, phenomena well described in terms of one of these metaphors may be the result of component processes well described in terms of another. Thus, response and ontogeny, where they occur, are often realized by means of lower-level impressions and sortings that themselves are ultimately realized by means of simple physical causes at the molecular level. (For example, the impressive capacity of a mammal's immune system to mount an active response to a particular microbial invader works by means of a process of selection [sorting] acting on the cells of the immune system, but this in turn is ultimately achieved by means of simple impressions at the chemical level—the change in a receptor caused by its target molecule binding to it, and that is simply caused by the electromagnetic forces linking the relevant molecules.) What matters for our purposes is which—if any—of these metaphors is successful in capturing significant aspects of the high-level patterns found in organismal development and behavior, and especially in human cognitive development and social behavior.

If human development and behavior mainly follow the pattern of conservative interaction, it seems plausible to say that although reformers might strive for substantial social changes, these are likely to come only at an unacceptable cost. In a world dominated by conservative interaction, large-scale changes in human behavior would seem to demand large-scale environmental interventions that would be difficult to generate and could be expected to have extensive additional effects, "forcing" human developmental and behavioral patterns beyond their normal (or even their normative) bounds.

If the pattern of radical interaction is common, however, the prospects for social change look quite different and far more hopeful. The metaphor of response provides a conceptual starting point from which to begin examining some theoretical and empirical results that support and fill in this alternative picture.

[4]

Lessons from Development, Ecology, and Evolutionary Biology

ORGANISMS RESPOND TO THEIR ENVIRONMENTS, but they also change them. This creates the possibility of a feedback loop—organisms respond to the environments they themselves have changed, and their responses then affect how they go on to change those environments. The importance of this sort of feedback for human societies has long been recognized. The ecological psychologist Roger G. Barker noted it back in 1963: "Who can doubt that changes in our environment ranging from new levels of radiation, to increased numbers of people, to new kinds of medicines and new kinds of social organizations, schools, and governments are inexorably changing our behavior, and that our new behavior is, in turn, altering our environment?" (1963, 19). More recent research clarifies how such feedback can play out, and shows just how consequential it can be. To understand the full implications of this causal entanglement between organism and environment, we need to examine both of the strands connecting them: the quality of plasticity that enables organisms to respond to environmental influences and the process of niche construction by which organisms modify their own environments—and each other's.

Plasticity

Important work emerging from developmental biology over the past two decades emphasizes the ubiquity and importance of phenotypic plasticity (Pigliucci 2001, 2005; Pigliucci, Murren, and Schlichting 2006; West-Eberhard 2003). The term is used broadly to talk about genotypes that can produce markedly different phenotypes in different environments. The metaphor of "plasticity" suggests malleability or direct impression, and these metaphors correspond well to what biologists call "passive plasticity." But many important instances—those that biologists describe as "active plasticity"—fit the metaphor of response much better. Whereas passive plasticity is a matter of mere vulnerability to environmental stresses and results in unregulated change in the organism, active plasticity is expressed in complex and coordinated change, often integrating adjustments in morphology, physiology, behavior, and even life history, regulated by means of complex multilevel feedback processes (Whitman and Agrawal 2009). Two birds with the same genotype might differ in color because one has lacked access to the nutrients that are needed to produce a certain pigment in its plumage—an instance of passive plasticity. Or they might differ because one has put on breeding plumage and the other has not, owing to their exposure to differing local seasonal cues—an instance of active plasticity. In the latter case—but not the former—the change in plumage color is just one part of a suite of coordinated changes. A male bird putting on breeding plumage will also adjust his social behavior (courting potential mates and threatening potential rivals, for example) and may build a nest or bower or begin signaling or patrolling to defend his territory. He will almost certainly change his song patterns and may develop brightly colored patches of exposed skin on his face or legs or even grow special plates that change the apparent shape and color of his bill. These visible changes—and the underlying physiological changes that make them possible, providing additional neurotransmitters, pigments, blood flow, and nutrients—are all controlled by hormone levels that themselves change in response to environmental triggers.

Developmental biologists also distinguish between adaptive and nonadaptive plasticity, where adaptive plasticity results from the operation of an evolved mechanism that allows the organism to tap different developmental or behavioral strategies under different environmental conditions, thus maximizing its fitness in a variable environment. This distinction is obviously related to the passive/active distinction: active plasticity is likely to be adaptive.

Phenotypic plasticity can be structured in various ways: it can be continuous or discontinuous, restricted to a brief developmental window or ongoing throughout the life of the organism, reversible or irreversible (Pigliucci 2001; West-Eberhard 2003; Whitman and Agrawal 2009). Some properties of organisms vary continuously with environmental factors: in many organisms, for example, size varies continuously with the temperature the organism experienced during its development (within certain bounds). Discontinuous variation can result when the organism possesses distinct alternative developmental pathways, separated by environmentally sensitive "switches": in alligators and some turtles, for example, sex is determined by the temperature in the nest early in the incubation process. A small difference in temperature can result in a different setting of the "switch" that determines whether the embryo will develop as female or male. This case is one of many in which divergent developmental pathways are accessible only at a particular developmental stage—temperature affects sexual differentiation only during a brief developmental window. There are also many cases, however, in which divergent pathways are accessible throughout the life of the organism. Clownfish begin life as males but can become female either early or later in life if no dominant female is present in the local environment. Some changes, like the clownfish's, are irreversible once they occur, but others (such as seasonal changes in mammals' coat color, birds' plumage, or plants' foliage) are readily reversible.

Most research on phenotypic plasticity focuses on morphological examples like these, but behavior is involved with plasticity in several different ways. Behavior itself—what Richard Dawkins called "the trick of rapid movement" (Dawkins 1976) could be seen as a form of very swift, highly reversible phenotypic plasticity. Larger patterns of behavior can be plastic at various scales or levels—they can be modified in various ways in response to environmental factors, as in psychological priming, imprinting, conditioning, and more sophisticated forms of learning. Developmental plasticity in animals can also involve aspects of neural and endocrine system development that may be developmentally locked in with lifelong effects on behavior. Rat pups neglected by their mothers, for example, grow up with more reactive cortisol systems, while rats raised in environments offering enriched sensory and cognitive stimulation grow up with thicker cerebral cortexes and denser neural networks; both changes continue to affect their behavior in adulthood. In other cases, such neural and endocrine changes may themselves remain plastic in varying degrees. Recent research on neuroplasticity, for example, has begun to reveal the remarkable ongoing adaptive capacity of many aspects of the brain's structure and function.

The evolution of plasticity has been a major focus of research in recent decades. Evolutionary theorists long discounted the importance of adaptive plasticity on the supposition that it was very difficult to achieve—requiring elaborately integrated mechanisms for detecting and responding to environmental conditions, which could evolve only in lineages exposed to reliably repeated environmental variations with distinct selective pressures—and so must be a rarity. But this conclusion has proven to be mistaken; adaptive plasticity is ubiquitous, as are the various particular adaptations that make it possible. In particular, nervous systems and endocrine systems—both crucial to human sex differentiation as well as to many of the other features of human nature emphasized by evolutionary psychologists—are adaptations *for* plasticity.

A few additional aspects of phenotypic plasticity are particularly worth noting. One is the way in which adaptive plastic responses are evoked. The environmental cue that triggers such a response is very often not identical with the environmental change to which the response is adaptive but is a proxy or indicator that the organism can detect more easily or earlier. This enables organisms to show anticipatory plasticity: adaptive changes that appear in advance of the environmental changes that select for them. Many seasonal changes in plants are adaptive to changes in temperature, precipitation, or pollinator presence, but plants often actually initiate such changes in response to changes in the amount of daylight, a (usually) reliable indicator of coming changes in the biologically significant variables. Often such changes are cued by the proxy variable's crossing a threshold, which means that small changes in proxy variables can sometimes trigger large changes in plastic characteristics or behaviors. In a famous case from horticulture, a large section of a greenhouse of poinsettia plants was prevented from blooming by the night watchman's habit of stopping at the same spot each night to light a cigarette: poinsettias begin their blooming cycle in response to the long nights presaging the rainy season in their home environment, and the brief flash of the watchman's lighter broke the night into two segments, neither of them long enough to trigger blooming. In other cases (where the threshold is not crossed) large changes in the proxy variable may have no effect at all on plastic phenotypes.

Finally, since proxies are important only because of their correlation with biologically significant variables, it is not always obvious what they are even if the adaptive function of the plastic response is known. The discovery that photoperiod-sensitive plants like poinsettias respond to seasonal changes in daylight was an important one, but the further discovery that they are responding to night length, rather than day length, took almost two more decades. Because animals have a far richer array of mechanisms for sensing environmental variation than plants do, identifying the proxies that animals use as cues can be very challenging.

Not only the proxies but some plastic responses themselves may be difficult to discover. Interesting recent work suggests that organisms may harbor considerable "latent" plasticity, some of it perhaps surviving from much earlier evolutionary periods—plasticity that is tapped very rarely if at all simply because the environmental triggers needed to elicit it occur rarely or never.

A final point worth special note is the complex relationship between phenotypic plasticity and phenotypic stability. Although plasticity is usually associated with change or variability, it can play a key role in producing stability or uniformity if the particular environmental factors to which a plastic response is linked are themselves stable or uniform. One important kind of case to consider here involves "scaffolds"—structures in the environment that serve as guides or templates for plastic developmental processes. The mature form of twining plants is determined by the structures about which they twine, for example, while the array of speech sounds that adult humans can distinguish in hearing is determined by the patterns of meaningfully different speech sounds they were exposed to as young children. The acquisition of particular concepts and bodily skills is similarly scaffolded by artifacts and cultural practices. The things we surround ourselves with—tools and toys, vehicles and weapons—shape our minds and capacities in ways that are sometimes passed from generation to generation for remarkably long spans of time (Sterelny 2012).

Plasticity can also play another role in producing stability by enabling an organism to respond differently to different external influences so as to reach or maintain a stable outcome despite environmental variability. The plastic response of individual branches allows a tree to arrive at the same overall mature form even if it loses limbs or must grow around obstacles; the plastic responses of the brain allow it to compensate for injury by recruiting other regions to perform functions normally executed by the damaged region. Most stability in living things is of this sort: robust rather than rigid; actively maintained and adjusted by plastic response in such a way as to compensate for perturbations rather than resulting from simple resistance to modification. Such robust stability is in fact essential to the operation of plasticity that is actively responsive rather than passively malleable. The alternative developmental pathways that endow an organism with active plasticity are themselves robustly structured, and their activation by environmental triggers is also reliably achieved despite other kinds of variability. Temperature-based sex determination offers a striking example. For a brief period, as we noted earlier, alligator embryos are highly sensitive to the temperature of their nests—which determines whether they will develop as males or females—but robust to other environmental variation. Once their commitment with regard to sex is made, however, the developmental process in either pathway is robust to further fluctuations in temperature. Another example is an equally striking case of conditional sexual strategies. Male Atlantic salmon have two different ways of reaching sexual maturity. Some fish grow to a large size (from fifty centimeters to a meter in length) and take a long time to reach maturity (up to seven years). Others mature at a very small size—as small as six centimeters—and reach maturity in less than half the time taken by the large fish. The large males (anadromous males) mature at sea, while the small ones (mature parr) are able to reach maturity without leaving the rivers where they were spawned. Which strategy a young salmon will follow is determined by his growth rate early in life—growth rates below a certain threshold trigger the process of early maturation. Different genotypes have somewhat different thresholds, but for each genotype the two developmental paths are robust once under way, even though tiny differences in early life can make the difference in which path a fish follows.

What are the implications of all this for thinking about human response functions? It is clear that sensitivity to environmental variability—especially to variability in social environments—and the capacity to respond plastically to it are enormously important to human beings. We are evolved for plastic response at many levels: developmental, behavioral, cognitive, and cultural. But much of the plasticity we possess is directed toward maintaining the stability of certain outcomes in the face of the enormously diverse physical, biotic, and social environments in which we find ourselves. Thinking about the role of plasticity makes clearer what we need to know in order to be able to draw conclusions about the shape of response functions. We need answers to these questions: Which phenotypic features are plastic, and what environmental proxies are they sensitive to and in what ways? Which features are robustly maintained and how? Uniformity of a particular characteristic does not show that it is not plastic since there are many reasons that available plasticity may not be in evidence. There is also reason to expect some apparently minor factors—proxies for biologically significant variables—to trigger strong responses. Some of these triggers (those affecting what I have called "biological levers" [Barker 2008]) may be capable of initiating causal cascades with far-reaching effects, as peoples' responses change their effects on others, triggering further responses and further environmental changes.

None of the ideas about plasticity just canvassed would be news to evolutionary psychologists. They have recognized from the beginning that both behavior and development are plastic and responsive to environmental variation, and in the last few years some have begun to place much stronger emphasis on the role of plasticity in the genesis of individual variation and in the production of human behavior (Brown et al. 2011; Buss 2009; Buss and Greiling 1999). As David Buss and David Schmitt put it, "Evolutionary psychology contends that human behavior is enormously flexible—a flexibility afforded by the large number of context-dependent evolved psychological adaptations that can be activated, combined, and sequenced to produce variable adaptive human behavior" (2011, 781).

Yet they also insist that the flexibility goes only so far. Two kinds of limits are often identified. First (evolutionary psychologists insist), though the surface behavior may vary, the underlying cognitive mechanisms remain the same. Second, though behaviors or phenotypes may vary in some domains, those at the adaptive core—those comprising human nature—are stable, for they are what the plasticity has evolved to protect. In the case of sex differences, this means that where the sexes face substantially different adaptive problems, the differences between the solutions that they have evolved will be substantial and relatively invariant. These claims give more specific substance to vague assertions about limits to human malleability, but they also raise questions. Are the lines separating what is plastic from what is fixed really as sharp as all that? Are we completely sure about where they fall? Chapter 5 examines these questions more closely.

Given the many levels of human plasticity, the emphasis that advocates of evolutionary psychology (and some of their critics) have often placed on the conflict between thinking of the mind as a "blank slate" and thinking of brains as "hardwired" for certain social patterns can be seen to be badly misplaced (Tooby and Cosmides 1992; Pinker 2002). What is wrong with the "blank slate" metaphor is not that it posits plasticity where there is none but that it misrepresents plasticity as malleability—mere susceptibility to external impression—rather than as response. What is wrong with the metaphor of "hardwiring" is not that it posits stability where there is none but that it misrepresents stability as mere rigidity rather than robustness. The question about human possibilities is not about the limits of human malleability and the extent of human rigidity; it is about the range of human responses—developmental, behavioral, and cognitive.

Niche Construction

Organisms change in response to their environments, but they also change those environments. Organisms modify their own and others' environments in ways that affect both selection pressures and ecological influences—a process that has been dubbed "niche construction" (Odling-Smee, Laland, and Feldman 2003). Some niche construction is adaptive—classic examples include the construction of nests, spider webs, and beaver dams. Other instances are merely side effects of adaptive processes, like the production of various wastes. Niche construction by one organism or population can affect other organisms or populations, and its effects can endure through generations, creating a channel of "ecological inheritance" that accompanies and interacts with the genetic inheritance system. Thus, for example, forest trees inherit not just their parents' genes but also the rich soil that earlier generations of trees have helped to create. Young beavers inherit both webbed feet and the flooded environments that make webbed feet so useful.

Niche construction can have powerful evolutionary impact, generating feedback loops between the features of organisms that help bring about environmental change (like the beaver's teeth and dam-building behavior), and those that are affected by the modified selection pressures that such change produces (the webbed feet that adapt it to a flooded environment). These feedbacks can lead to evolutionary effects such as rapid evolution, the fixation of otherwise unexpected traits, and the establishment of associations between traits (Day, Laland, and Odling-Smee 2003; Laland 2006; Laland, Odling-Smee, and Feldman 2009; Robertson 1991). The evolution of web spinning in spiders, to take another example, created a host of new selection pressures resulting in a remarkable array of new behavioral and morphological adaptations to living on spider webs, which in turn enabled the evolution of new elaborations of web construction.

Humans are niche constructors par excellence. In all human societies, people live in environments modified by human activity (at least on a local scale) in ways that affect our physical and cognitive development as well as our behavior, welfare, and prospects for survival. We engage in physical niche construction, building fires and weapons, roads and shelters. We engage in biotic niche construction, domesticating and modifying plant and animal species, clearing forests, and planting fields. And, at least as importantly, we engage in social niche construction, creating marriages and battlefields, parliaments and schools, slave markets and chess clubs (Dean et al. 2014; Kumm, Laland, and Feldman 1994; Laland, Odling-Smee, and Feldman 2000; Mesoudi and Laland 2007; Rendell, Fogarty, and Laland 2011). Niche construction is crucial for human thriving, and its importance creates additional selection pressures favoring plastic response—niche-constructing behaviors need to be tuned to changing environmental conditions while behavior in general needs to be tuned to the constructed environments that may change repeatedly over long human lifespans (Sterelny 2007, 2011). On the other hand, our niche construction often contributes to the maintenance of robust phenotypic stability because reliably reproduced features of our constructed environments serve as an ecological inheritance system that supports the reproduction of phenotypic traits across generations (Sterelny 2003, 2007, 2011, 2012). The tender feet shared by many modern humans result from the shoes we make and wear; our ability to read and to drive—robustly recreated across generations—results in part from growing up in environments containing texts and vehicles.

In humans and in other species, niche construction can also contribute to the stable maintenance of discontinuous variation, including sexual and other polyphenisms—cases in which organisms of the same species develop along two or more quite different pathways, as do the sexes in many animal species, or the different castes in some species of social insects. Plasticity, as we noted, can involve a developmental "switch" (producing discontinuous variation as in alligators' temperature-based sex determination) or a graded response to an environmental variable (producing continuous variation—the influence of temperature on size determination in many insects works this way). Recent research suggests, however, that many cases of discontinuous variation in phenotypes are actually the result of graded responses to discontinuous environmental variation where the discontinuity is in turn maintained by niche construction. In honey bees, for example, the dramatic difference between worker and queen bee morphs looks like the effect of a developmental "switch" like those involved in sex determination. But this difference has been found to be produced instead by discontinuous differences in the food provided to larvae. Nurse bees feed most larvae a worker diet while they feed selected larvae the sharply different diet that creates queens. If scientists feed bee larvae a diet intermediate between the two, bees develop adult body forms intermediate between the two normal morphs. Discontinuous variation in human traits is also sometimes caused by discontinuous environmental variation that is maintained by social niche construction. Thus, for example, many sex differences are exaggerated or created by environmental interventions, whether deliberate or not. (Deliberate cases include ornamental body modifications; cases that are not deliberate include at least some of the developmental differences caused in some populations by differential nourishment, physical treatment, clothing, and education provided to young girls and boys.) Such differences can also separate members of different social classes—archaeologists have found discontinuous variation between landowners and serfs or slaves in some historical communities, owing to sharp differences in their nutrition (Boix and Rosenbluth 2007).

Human niche construction is unique in combining social and physical niche construction to produce what some ecological psychologists call behavior settings (Barker 1968; Schoggen, Barker, and Fox 1989; Heft 2001). These are specialized local environments constituted by material objects and human participants interacting in well-defined ways: settings such as classrooms, kitchens, highways, basketball courts, and courts of justice. Behavior settings are powerful shapers of human behavior, and they can be robustly self-perpetuating and self-correcting. They can sometimes be reliably maintained or reproduced across many human generations and can thus play an important role in reinforcing existing patterns of behavior and development. But they can also facilitate change. New behavior settings can be created, for example, when new technologies or activities are introduced and can establish and stabilize new patterns of behavior; old behavior settings can also facilitate change when they are recruited to new uses. Women's progress toward social equality in the late nineteenth and early twentieth century was assisted by new behavior settings that emerged around new technologies, such as bicycle clubs and textile mills. But it was assisted as well by the stability of existing behavior settings—such as university classrooms—to which women were winning new admission.

The implications of these aspects of niche construction for the question of human possibilities are obvious. Our bodies and brains develop in and respond to environments that we actively construct and maintain; they have evolved to do so. Some very stable and widespread human features are maintained by means of robust niche construction (obvious examples include the nutritional effects of using tools to augment our capacity to hunt or farm and using heat to cook food, and the cognitive effects of using language in interacting with babies and children). Some locally stable polyphenisms—such as certain differences in phenotype between the sexes or between social classes—are likewise maintained by the stable reproduction of environmental differences. The explanatory emphasis thus shifts from the internalist's question—"What kinds of development and behavior are humans (or humans of a particular sex) prone to express?"—to a more complex one—"What kinds of environments are humans prone to create and sustain under what conditions, and how do humans respond (both developmentally and behaviorally) to those environments?"

[5]

Human Possibilities

THE IDEA THAT HUMAN NATURE PLACES easily identified limits on the kinds of societies we can hope to build turns out to be founded on a broad assumption about how internal and external factors—genes and environment—interact to shape human behavior. If the interaction follows the pattern I describe under the rubric of "conservative interactionism," then claims about the limits set by human nature are realistic. If the pattern is closer to that of radical interactionism, the question becomes more complicated, but the prospects for substantial social change are surely better.

Recent work on how organisms respond to and modify their own environments—the plasticity and niche construction discussed in the last chapter—casts new light on the key question of how internal and external factors interact in human development and behavior. Before returning to the question of what is possible for humans, it is worth pausing to consider some points about how internal and external factors interact to produce the actual patterns of human development and behavior seen across various environments and how plasticity and niche construction contribute to these patterns.

Some phenotypic traits in humans seem to conform fairly well to the pattern expected from conservative interaction while others are much closer to that expected from radical interaction. Eye color is a good case of the former type—environmental factors such as nutritional deficiencies may change eye color slightly by limiting the production of pigments, and extreme environmental variation may entirely prevent correct development of the eyes, but across a wide range of environments a given genotype will produce fairly consistent eye color. Immunity is a good instance of the latter type of trait—two people with the same genotype may have radically different antibody profiles as a result of their differing histories of antigen exposure. It is easy to see why these two broad patterns should be common ones: organisms need to keep certain traits stable in the face of environmental or genetic variability and need to adjust others in order to compensate for such variation. Evolutionary psychologists think that the features they identify as aspects of human nature fall into the first category. But it is important to be careful here.

First, we need to keep in mind the basic point that the patterns we see in human behavior are not full response functions—many possible environmental circumstances are missing from our current samples; many (a great many, indeed!) have never been realized at all. A second key point is that those patterns do not straightforwardly reveal the structure of the underlying causes. Often the partial pattern we see resembles what would be expected from conservative interaction as captured in the metaphors of limited malleability or constrained unfolding. But what we have learned about plasticity indicates that this pattern is not a simple reflection of the mechanisms of change but a complex achievement. Robustness and responsive sensitivity in developmental and behavioral outcomes look like opposites, but both are achieved by means of plastic response to environmental variation at some level of organization. Indeed, robustness and sensitivity are often found within different parts of the same response function. Even very sensitive response functions are often robust across certain environmental ranges, and even very stable phenotypes are often sensitive to some kinds of environmental variation. Discontinuous plasticity of the form in which alternative robust developmental paths are separated by sensitive "switches" (as we saw in the cases of temperature-based sex determination in some reptiles) is one common manifestation of such a combination.

This reinterpretation of the relative stability of outcome seen in conservative interaction has important implications for thinking about the shapes of human response functions and, in particular, for projecting from the parts of the response functions that we have seen to the parts that are still unknown, including the large stretches corresponding to untested social environments. The robustness of a phenotypic or behavioral outcome across current environments is not conclusive evidence for its "limited malleability" but is compatible with the supposition that the actual outcomes that we see represent only a limited segment of a much broader range of possibilities, perhaps corresponding to only one of two or more quite distinct strategies that the organism can adopt in responding to different environments. Without knowing a good deal about the particular developmental processes involved, it is very difficult entirely to rule out the possibility of such unrealized potentials or latent strategies. Examples of these potentials sometimes emerge when organisms are exposed to unusual environments. Animals living in wildlife sanctuaries, for example, sometimes strike up surprising cross-species friendships in which they may express behavior that is never seen in individuals living in normal wild settings. In one famous case noted by the philosopher and cognitive neuroscientist Patricia Churchland, an orangutan became very attached to a stray dog—the two spent all their time together, playing and resting together and showing great trust and mutual affection. Adult orangutans are normally solitary animals that do not form extended social bonds—certainly not with members of other species—and dogs are normally wary of primates, yet the potential for unexpected behavior by both animals was unlocked by the exceptional environment of the sanctuary.

What does this point mean for discussions about human possibilities? When evolutionary psychologists insist that human nature cannot be overridden or escaped, they are asserting that, with regard to the core features of human nature, we can indeed be sure that phenotypic and behavioral plasticity are insufficient to enable human populations to produce actual patterns of phenotype and behavior much different from those we see realized today, except in environments so extreme that their other effects would be unacceptable. So they are asserting that relevant alternative strategies—strategies capable of producing the kinds of outcomes desired by "utopians" in environments that could be easily and harmlessly achieved—are not now being expressed and are not latently available. They support these assertions partly by appeal to data about modern human populations and partly by appeal to assumptions about the evolutionary basis of human nature. Both of these lines of argument can be challenged, however.

The first argument appeals to a large and rapidly growing array of studies of modern human populations (Schmitt 2005; Schmitt et al. 2008) showing that some of the core behavior patterns expected by evolutionary psychologists hold across a wide array of populations in different countries, classes, or cultures. On the face of it, this evidence is very impressive. The question to ask, however, is whether these studies really show that the relevant kinds of plasticity are not now in evidence in these populations. Evolutionary psychologists emphasize the consistency with which the patterns that they identify with "human nature"—usually in the form of trait averages—hold across different cultural contexts, but they give relatively little attention to the range of individual variation *within* populations that those patterns also incorporate.

So researchers might compare the average number of sexual partners that men and women report in different cultures or the average score of girls and boys on a test of mathematical ability, but they do not closely examine the range of diversity within each group. If we are in the business of identifying what is "natural" to humans, though, we need to attend to the full range of that diversity, not just variation in average trait values.

Explaining Diversity

Optimism about human possibilities has seemed plausible in large part simply because the known range of human variation includes a substantial minority of individuals who show traits and behaviors very much like those that optimists would wish to see predominate: individuals who are unusually peaceable, generous, and egalitarian; individuals who readily form close relationships across cultural or racial boundaries; women who possess some stereotypically masculine strengths and virtues and men who possess some stereotypically feminine ones. But evolutionary psychologists have only recently begun to be seriously interested in individual variation (Buss 2009; Buss and Greiling 1999; Perilloux et al. 2010). Such variation, indeed, was long dismissed by some leading theorists (Tooby and Cosmides 1990) as mere "noise," irrelevant to any understanding of the "species-typical" traits that are the main concern of evolutionary psychology.

Certain kinds of explanations for some of the non-species-typical traits that optimists admire have nonetheless been central to evolutionary psychology since its early days. Some such variants are explained as strategies that can persist only at low frequency in a population: thus, game theoretic models show that a peace-loving "dove" strategy may do well if it is a rare variant in a population composed mostly of aggressive "hawks," but that a population entirely composed of "doves" is unstable and will inevitably be invaded by more hawkish individuals.[1] Such explanations usually assume that the different strategies are realized at the genetic level so that "hawks" and "doves" have different genotypes, though more sophisticated versions assume instead that there are genetic differences that confer differing tendencies in how often to "play" each of the two strategies. Other traits are understood as the outcome of the "misfiring" of evolved mechanisms in environments that are unlike those in which the mechanisms evolved. Thus, one purported explanation of male homosexuality sees a cluster of atypical male traits as resulting from exposure in utero to abnormally low levels of testosterone—traits that evolved as adaptations in females are "erroneously" triggered in male fetuses by this abnormal pattern of hormone exposure, despite the fact that they may (as those who favor this sort of explanation often suppose) be maladaptive for the male individuals affected. The existence of women (whether lesbians or not) with stereotypically male talents and interests has similarly been explained as the result of exposure to unusually high levels of testosterone in the womb.

These ways of understanding diversity may have encouraged evolutionary psychologists to assume that existing variation within human populations is no indication that desirable phenotypes that are now exceptional could become the norm without costly intervention, for low-frequency genetic strategies are (by assumption) available only in a minority of individuals, and misfiring is provoked only by markedly abnormal environments. But note the danger of equivocation between descriptive and normative notions of "abnormality" and "misfiring" here. Some of the environments thought to elicit misfiring are statistically common today—they are abnormal only in that they differ from ancestral environments. Our tendency to overindulge in sweet, salty, or fatty foods; our fondness for baby-like puppies and kittens; and our tendency to see human moods in our machines (spirited sports cars, uncooperative computers) are all misfiring responses to environmental features that are now ubiquitous in some societies.

The notion of misfiring in this context is supposed to be a technical and descriptive one: an adaptation misfires if it produces an effect that tends to reduce fitness. But examples such as the explanations of differences in sexual orientation and gender characteristics just mentioned show how easy it is to slide from this sense of misfiring to a more normative one, one in which atypical individuals are seen as the outcome of harmful developmental mishaps or errors. If we assume that atypical outcomes must come at a cost for the individuals involved in the sense we encountered in chapter 2, we are making a mistake—even if we restrict the assumption to outcomes that reduce the fitness of the individuals affected. Indeed, the ease with which some kinds of misfiring can be provoked should be seen as indicating the surprising extent of change that can be harmlessly effected by modifying human environments. Modern humans' altruistic behavior toward non-kin, for example, has been explained as resulting from our close interaction with non-kin at life stages when our ancestors would have interacted only with close relatives, so that mechanisms of imprinting that evolved to support altruism toward kin are "erroneously" triggered in interactions with non-kin in modern environments. More recent thinking in evolutionary psychology, moreover, has begun to consider other explanations of the broad range of both heritable and nonheritable individual variation in humans—including diversity in sexual orientation and gender roles. Some diversity is explained by plastic developmental and behavioral strategies capable of producing markedly different adaptive outcomes in different environments (Penke 2010), and some may be explained by recent divergent evolution of local populations under diverse local selection pressures. We'll look at these ideas in greater detail shortly; for now my point is simply that these new perspectives on human variation within evolutionary psychology reveal it to be far more open-ended than earlier evolutionary psychologists had often assumed.

The second evolutionary argument for the supposition that the core features of human nature are among those robustly maintained in the face of environmental variation is based on a presumption about our evolutionary history: that our ancestors faced certain vital problems so consistently over the relevant span of evolutionary history that the strategies that best solved those problems have been passed down to us as inflexible or obligate features. The environment that posed these defining problems of human evolution—the environment of evolutionary adaptedness (EEA)—is taken to be that typical of environments our ancestors occupied during the Pleistocene epoch, so the claim can be made more specific: No strategies providing conditional alternatives to those comprising "human nature" are now latently available to us because none were selected for in the Pleistocene. But this supposition can be challenged on several grounds. Research on plasticity in other species suggests that developmental capacities can survive for remarkably long evolutionary spans if they remain almost entirely latent and thus protected from strong selective pressures (Boomsma and Nygaard 2012; Rajakumar et al. 2012; Tomic and Meyer-Rochow 2011; Walia et al. 2010). This suggests that humans may retain latent capacities that evolved in much earlier evolutionary periods if the environmental conditions that activate them were sufficiently rare in the Pleistocene to prevent selective elimination and have remained so since. The Pleistocene itself is likely to have been quite variable in the local physical and social environments it presented to our ancestors. To take a simple example not often mentioned in descriptions of the EEA, both women and men undoubtedly sometimes found themselves needing to get along in the world and raise their children without the support of a mate.

Once the possibility of plastic developmental and behavioral strategies is admitted, it becomes apparent that rare but recurrent and significant selective problems are not necessarily overwhelmed by more common ones; instead they may result in conditional strategies with highly specific environmental triggers. Again, research on nonhuman animals suggests that quite rare environmental conditions can enable the evolution or maintenance of an alternative strategy if the distinctive selection pressures they bring to bear are strong enough.

Finally, as just noted, the presumption that human evolution has achieved little since the Pleistocene is subject to increasingly broad challenges (Hawks et al. 2007). If recent human evolution has been rapid, the diverse social and physical environments that humans have occupied in the Holocene (since the end of the Pleistocene, about twelve thousand years ago) may have produced a correspondingly broad range of environmentally cued behavioral capacities. What all this means is that there are no strong grounds for presuming that the range of human response to environmental variation must be limited to those that would have been most advantageous in the most common social and physical environments of the Pleistocene.

A little reflection on the role of niche construction in producing both stability and diversity in phenotypic and behavioral outcomes reveals further reasons to reconsider the range of possible human variation. On the one hand, the complex causal pathways by which humans achieve stable phenotypic and behavioral outcomes despite environmental variation are often not entirely internal to the individuals involved—they are commonly routed through constructed niches (material and social) in a way that may mean that these outcomes can be changed by modifying the niches. We create consistent local environments for ourselves and our children, and those environments help keep our behavior consistent too. But what if we tried changing the kind of environment we create? Evolutionary psychologists have written dismissively about attempts to do this by "utopian" kibbutzniks and hippy commune dwellers. But could a better understanding of social psychology allow us to create more successful alternative social niches? I return to this question in chapter 8. The success of any such venture would depend on the possibility of feedback between changing environments and changing behavior. Some of the environmental variation that triggers plastic responses in human behavior is created by human activity—activity that is in some cases itself an expression of plastic adaptive strategies. The fact that this sort of feedback can occur suggests an interesting possibility: certain small environmental changes may be able to trigger cascading effects by modifying the activities by which humans shape their own environments.

These arguments are abstract, and it is not easy to see what they have to teach us about actual human possibilities. We'll look at some concrete—indeed, practical—examples of what these ideas look like in action in later chapters, but it will be useful first to see how they might reorient one of the cornerstones of evolutionary psychology, the theory of sexual strategies.

Sexual Strategies

The behavioral ecologist Patricia Gowaty has developed a new view of sexual strategies that provides a striking illustration of what phenotypic plasticity and niche construction might imply for the question of human nature. Her basic idea is that human sexual strategies themselves (for both sexes) may be flexible: capable of adjusting adaptively to different environments. What environmental differences would require adjustment of mating preferences and behaviors? Gowaty argues that the most important variable in human social environments, for the purposes of sexual strategy, is how much reproductive autonomy women possess.

The conventional view of sexual strategies (sketched in chapter 1) is based on the presumption that men vary much more widely in reproductive success than women do. This is because men's reproductive output is limited primarily by how many times they mate, and the most successful males can win far more mating opportunities than their competitors. Women's reproductive output, by contrast, is limited primarily by the (relatively invariant) number of children they can bear and raise. Men have been therefore selected for traits that tend to increase success in intense competition for mates, while women have been selected for traits contributing to success in raising offspring to healthy maturity as well as traits that attract high-quality mates who will contribute to the offspring's success by providing good genes or a generous investment of resources. The divergent sexual strategies that result are said to be responsible for the differing characteristics that evolutionary psychologists see as typical of the sexes. For males, recall, these include aggressiveness, ambition, promiscuity, risk acceptance, interest in and competence with physical systems, and high variance in intellectual capacity ("more geniuses; more morons"); for females, the characteristics include empathy, domesticity, sexual choosiness, risk aversion, interest in and competence with social relationships, and low variance in intellectual capacity ("more mediocrity"). According to this picture, these evolved sexual strategies (and the differences between them) make up an inflexible part of human nature and constrain human possibilities in obvious ways. As Michael Ruse put it, "women want to spend time with their young children in ways that men do not" (Ruse 2012, 196) and men likewise care about—and have talent for—the pursuit of political power, mathematical understanding, and diversity in their sexual partners in ways that women simply do not.

The standard story was challenged at the outset by critics who pointed out that not all males would be well served by the strategy of trying relentlessly to rack up more matings. The philosopher Philip Kitcher argued, for example, that the supposedly fierce male–male competition for mating opportunities would in fact favor a conditional strategy for males (Kitcher 1985). A male who assesses his own chances of obtaining matings with females other than his partner as being quite small will do better to stay with that partner, while one who rates his own chances highly may indeed do better to play the field. Does this point require a serious revision to the patterns of behavior we should expect? That depends on how many of our male ancestors found themselves in the situation of having relatively ready access to new mates. If most of them did, the revision may be modest; if relatively few of them did, the revision becomes much more important.

Gowaty's criticism is related to Kitcher's but reveals even more fully how the possibility of conditional strategies strikes at the root of the standard story. She focuses on one crucial assumption upon which that whole story turns. The assumption is that women will agree in their mate preferences so that, when women choose their preferred mates, the result will be that the men they all favor—the "best" males, from a mate's point of view—end up with exceptionally high reproductive success. Gowaty argues that this key assumption is mistaken. She finds evidence instead to support a hypothesis implying that different women should prefer different mates.

According to this hypothesis, a key variable for female reproductive success in many animal species is the viability of offspring early in life—what proportion of the young they produce live to grow up. This is strongly affected by immunities that the offspring inherit from their parents, and those with genetically disparate parents inherit a wider set of immunities. So, the hypothesis goes, females' mate preferences have been tuned by natural selection to pick out mates with genes complementary to their own. There is good evidence that this is true for many animals, and some suggestive data for humans (Anderson, Kim, and Gowaty 2007; Botero and Rubenstein 2012; Roberts and Little 2008). For humans, this means that when women are able to choose their reproductive partners freely, each woman should have mate preferences as unique as her own genome. And this in turn means that men's variance in reproductive success will be comparable to women's after all, which means that the best sexual strategies for the two sexes will be much more similar than is usually thought.

On the other hand, when women's mate choices are constrained either by direct social intervention (such as arranged marriage or coerced mating) or by male control of essential resources, men are indeed likely to have much greater variance in reproductive success than women do and should favor a sexual strategy like that posited by mainstream evolutionary psychology. Under these conditions, women must pursue a strategy that helps them to compensate for the reduced viability of their offspring. They will do this by producing more offspring and investing more effort in raising them than would otherwise be optimal as well as by competing for generically good genes and reproductive investment from high-status mates. If women lack reproductive autonomy, according to this hypothesis, the sexual strategies selectively favored in women and men diverge sharply, resulting in differences very much like those that evolutionary psychologists see as an inescapable aspect of human nature.

Gowaty argues that both women and men have thus evolved flexible strategies that allow them to adjust to varying social environments, and that many of the supposed "hardwired" sex differences emphasized by evolutionary psychologists are actually facultative responses to social environments in which women's autonomy is suppressed either directly or indirectly. It is important to note that, if this picture is correct, some aspects of both male and female strategies may remain plastic throughout the life cycle, but other aspects may be expressed in irreversible developmental "choices" and therefore represent lifelong commitments. If early endocrine and brain development are responsive to features of the environment that serve as proxies for female reproductive autonomy (or its absence), the results may be irreversible. The question is what environmental indicators girls and boys (and women and men) might respond to in "choosing" their sexual strategies and how and at what stages of development the response takes hold. One crucial stage could be during fetal development, when the developing brain is exposed to a hormonal environment partly determined by the mother's responses to her own life circumstances. A further twist could well be added by later parental niche construction: parents may adjust their interactions with daughters and sons, including the behavioral scaffolding they supply, in response to environmental cues that are detectable by the parents but not directly by the children. If these possibilities are correct, we will not know the full behavioral effects of recent changes to the social environment for several decades.

The hypothesis that Gowaty presents gives a distinctive picture of human possibilities in the domain of sex differences—a model of sex differences as reflecting active responses by both sexes to environmental variation. This "facultative response" model fits the pattern of radical interaction between genes and environment. It resembles the externalist "impression" model of environmental influence in certain ways but also differs from it importantly. Like the impression model, Gowaty's response model implies that we cannot know what the possibilities are until we have actually surveyed all the relevant environments and the outcomes they produce—in particular, that we cannot know what female and male humans are capable of until we have seen what they do when born and raised in environments that do not act unequally to foster or constrain their capacities. It agrees, too, that substantial changes in phenotypes and behavior can be brought about by adjusting environments.

When we look more closely at the details of what the two models imply, however, the differences are striking. What kinds of environmental factors are likely to be important, and what kinds of effects are they likely to produce? According to the response model, the environmental factors that cue responses are merely proxies for factors that affect fitness. This means that the most effective environmental adjustments may well be less than obvious—indeed, perhaps apparently quite trivial—and that their effects may be surprising in their magnitude or direction. Many environmental changes, on the other hand—including quite substantial ones that seem intuitively relevant to behavioral sex differences—can be expected to have little effect at all since they do not affect the key variables to which responses are attuned. If Gowaty is correct, then indicators of female sexual autonomy should be critical, but we do not know what indicators different categories of people (pregnant women, babies, girls, women, boys, men) might use as proxies for this variable. The proxies could be quite diverse, and some of them might have only imperfect links to autonomy in contemporary societies. And because some environmental effects may be realized in very early development or even via epigenetic effects that extend over several generations, the facultative response model expects substantial and variable time lags in their manifestation. On the other hand, it also expects some very rapid responses that permit people to adjust their behavior fluidly to rapidly changing contextual cues.

Kitcher's earlier critique of the standard story still applies within Gowaty's model and adds a further and important twist. Even when female autonomy is (apparently) low and the classic male strategy is favored, it will be favored only for those males who deem their chances of multiple matings to be good. But what environmental features do boys and men use to assess their mating prospects? And what features do boys' parents use to assess the future mating prospects of their sons? Doubtless some general indicators of social status or "good genes" are important here, but other factors may play a significant role. It might be, for example, that exposure to a steady stream of images of (apparently) sexually available women (in advertising, entertainment, and pornography) serves as a misleading indicator of mating prospects for many men.

Gowaty's response model suggests that some kinds of interventions that people commonly use in trying to reduce sex differences in the choices people make in their education and work lives, for example, are likely to be ineffective. Relatively large environmental changes that do not involve any significant indicator variables will probably have little impact (encouraging children to play with toys stereotypically assigned to the other gender, for example). In general, the response model expects young people to respond more strongly to factors in the local population at large (where mating patterns are decided) than to those within their immediate families. Thus parenting choices alone may be quite ineffective, and choices that parents make unconsciously—perhaps in response to their own experience of the local social environment—may be far more powerful than conscious choices.

Feminists have argued that "the personal is political." Gowaty's hypothesis about reproductive strategies thus suggests that it is at least possible that the personal is more than political; it is biological. What might make the biggest difference in producing gender equality is ensuring that women have real freedom to choose their own reproductive partners and that everyone knows it. But the response model also suggests a new perspective on the challenges facing any effort to achieve this outcome. Where mainstream evolutionary psychologists predict that aspects of human nature that are universally shared will simply resist certain kinds of change, the response model expects active, complex, and ongoing social conflict over what kinds of environments we create and maintain for ourselves, our children, and each other. We have evolved complex strategies of response to changing conditions not only in our own development and behavior but also in the ways in which we act to shape the phenotypes and behavior of others—we are evolved not just to be responders but to be responsive niche constructors for ourselves and for each other. This insight has important implications for understanding what is going on in existing environments today. The recent rise of a popular "rape culture," "slut-shaming," and ambitious political efforts to limit women's access to abortion and contraception can be seen as responses to indicators of women's increasing sexual autonomy—responses whose adaptive function is to limit that autonomy—by men whose subjective sense of their mating prospects suggests that those prospects are likely to be diminished when women have more reproductive control. Note that the unconscious assessments that these men are making need not be accurate—here again, pornography and popular culture may give men misleading indicators of their own sexual prospects. Feminists' investment of effort aimed at defending and extending women's reproductive autonomy, by the same token, can be seen as an expression of an evolved and adaptive response to indicators of constraints on that autonomy rather than (as evolutionary psychologists have sometimes suggested) a temporary and unnatural aberration.

Recognizing the pervasiveness of our evolved tendencies to seek to control our own and each other's environments and the conflict that they produce has important implications as well for how to think about choosing environments for future generations.

[6]

Valuing Change

WHAT ENVIRONMENTS SHOULD WE SEEK to provide for rising generations and for those yet to come? How much should we be willing to invest in creating and sustaining particular kinds of environments? These are intensely practical questions whose answers have implications for our private choices but also for public policy about everything from education to urban design, and from medicine to policing. They are also frankly normative. Can an evolutionary understanding of human psychology help us to answer them?

So far this book has focused on questions about how—and how much—environmental changes could change human patterns of behavior. We've seen that these patterns may be more open to environmentally triggered change than most evolutionary psychologists have thought. But this alone means relatively little. We need to know not just what the effects of particular environmental changes would be but which of the possible effects are actually desirable and which of the required environmental changes are actually feasible. The first is a question about values, the second about facts. These interlinked questions are the focus of the rest of this book.

To assess the prospects for environment-driven social change, we need to think about both what we *can* do and what we *should* do—about what modifications to current patterns of human development, behavior, and social organization are feasible and what paths toward such change—if any—are good to pursue. Each of these considerations raises numerous finer-grained questions. On the factual side, to assess feasibility we would need to know what changes can be achieved at all, and which among these can be sustained once achieved. By what means can they be thus realized, and with what side effects? What are the likely outcomes should attempts to realize them fail? On the value side, we would need to know what makes an environment or a developmental or social outcome good or bad. Are there objective standards for these valuations, or do the standards vary with context or according to the attitude of those affected? How are diverse and uncertain goods and harms, unevenly distributed across the human population, to be weighed against one another?

Despite the complexity of these questions, evolutionary psychologists often write as if the answers can be read off fairly simply from evolved human nature. Their understanding of the interaction between organism and environment leads them to expect that human psychology will follow the patterns of conservative interactions, the broadly internalist perspective captured by the metaphor of "limited malleability." From this perspective, the facts and values in play here appear closely intertwined. Human nature is seen as determining the main patterns of individual development and behavior; these in turn fix what is feasible at the social level. Some social arrangements that are easy to imagine cannot be put into practice simply because human nature resists them: you cannot make enough people behave unnaturally enough, consistently enough, to make them work. Recall Robert Wright's remark that humans "aren't malleable enough to make communism a productive economic system, and they aren't malleable enough to create a society of perfect behavioral symmetry between men and women" (Wright 1994a). Evolutionary psychologists argue that American countercultural communes generally failed and Israeli kibbutzim were forced to abandon their most radical aims because they sought to realize imaginable but unfeasible social arrangements, including communal ownership of property, collective child rearing, and the elimination of traditional gender roles. This perspective sees only certain kinds of social arrangements as feasible—those that are compatible with the relatively unimpeded expression of human nature.

It might seem, further, that environments that permit human nature to express itself most fully are not only feasible but *good*: environments in which people are free—and happy—to behave as they naturally prefer. Evolutionary thinkers do often write as if these "natural" outcomes and environments are best for us and are the ones we naturally tend to create or re-create. They recognize (of course) that modern environments differ markedly in certain respects from the Pleistocene environment to which our nature fits us, and they trace some contemporary problems to the mismatch between our "stone-age minds" and the modern world. But in general (evolutionary psychologists seem to think) the environments we most readily create are good for us—they are themselves expressions of our nature, and they enable our development to follow its natural course (Veenhoven 2010).

On the other hand, evolutionary psychologists recognize the naturalistic fallacy as an obvious error, so they cannot simply assume that whatever is natural is good. Indeed, most make it clear that they see some aspects of evolved human nature as undesirable or morally problematic and think that environments that bring these to the fore should be modified—even if they are among the environments we most readily create. Richard Dawkins, for example, has worked with great dedication to help make our social environment more conducive to secularism (Dawkins 2006), and Steven Pinker has recently devoted a lengthy book to the pacifying and civilizing influences of factors such as central government, humanism, and the printing press (Pinker 2011). There is also, plainly, abundant evidence that some social systems that are widely agreed to be very bad for their members are achievable and may be sustainable over a period of generations at least. The simple idea that the environments most natural—and therefore most feasible—for us must also be good for us is clearly inadequate.

The "response" perspective developed in the last chapters offers quite a different view of the feasibility and value of pursuing social change by modifying environments. On this view it seems likely that humans have evolved multiple adaptive strategies of development and behavior that can be switched on or off in response to particular environmental cues so that human nature does not pick out one sort of developmental trajectory or outcome or one sort of environment as "natural." This active plasticity means that some novel social arrangements may be quite achievable—it may be possible to shift the behavioral patterns or even the developmental patterns of large numbers of individuals at the same time by adjusting key environmental cues. The role of niche construction means that some novel social systems may be able to reinforce themselves by reproducing their own enabling environments and so be sustainable. But where does this leave the question of value? Suppose that humans are capable of responding adaptively to a diverse variety of environments, environments that we are also capable of actively re-creating. Do we have any grounds for choosing one of these environments and the social arrangements that it incorporates over another equally feasible combination?

The response perspective does indeed imply that diverse phenotypic outcomes may sometimes represent adaptive responses to different environments equally "good" for different purposes or for achieving the same purpose in different contexts. But the scare quotes around "good" here matter: this does not mean that there are no important evaluations to be made here, merely that simple evolutionary considerations cannot do the evaluative job for us.

From an evolutionary point of view, all that matters is our success in getting our genes into the next generation, but much more than that matters *to us*. Some environments really do prevent people from realizing their potential—not just their reproductive potential but their human potential—and some phenotypes really do reflect this.

But humans have many potentials: Which should we consider? People disagree about which potentials matter, and even if we could somehow agree upon a list, we would find (as dystopian storytellers have long recognized) that different environments maximize different items on that list. Some of the human qualities, capacities, experiences, and achievements that we most value may prove to result from strategic responses to environmental challenges—responses that may also have effects that we regret. Worries about such phenotypic or behavioral trade-offs are surely well founded, but it is far from clear how to assess the possibilities that they present.

Evaluation: What Makes a Social Change Good?

How to make sense of the idea that some environments, phenotypes, and social arrangements are better than others is a problem for any attempt to understand human lives and choices in scientific terms, as part of the natural world. Whether or not you accept the idea that facts and values are sharply and deeply distinct, the place of value in a naturalist's universe is a problem. As we just saw, the complexion of the problem depends on our understanding of the respective roles of internal and external causes in development and behavior—whether we take the "limited malleability" or the "response" perspective on human nature—but in any case it remains to be solved. A further problem is more practical: Assuming that some answer can be given to the broad question of what value can be in a natural world, we need to know how to assess the relative value—for us—of particular ways the world might be. What exactly should we count, and how, in comparing possible environments and their effects on human lives? The focus in what follows is on this more practical problem and, in particular, on what a revised view of human evolution, development, and behavior can teach us about it.

As noted in chapter 1, when thinkers who seek to draw social lessons from evolutionary psychology point to "the cost of change," they are invoking an approach to the assessment of alternative options that has come to dominate debates over social policy in recent decades: cost–benefit analysis. None of these thinkers has explicitly endorsed cost–benefit analysis as a method for deciding what kinds of environmental intervention we should attempt, and none has tried to offer a systematic cost–benefit accounting of any particular social policy proposal, but the language that they use and the costs that they discuss point clearly to this well-known conceptual framework. This is not surprising: cost–benefit analysis is commonly used to assess alternative strategies in business and government projects, and it is also common in evolutionary theory for assessing the effects of different traits with their various trade-offs. Its apparent simplicity and comprehensiveness lend it a seductive appeal for those evaluating complex choices. Our first step in tackling the problem of value will be to examine evolutionary psychologists' ideas about the prospects for social change with an eye to the implications—and limitations—of this framework for social decision making and of the interpretations that they bring to it.

Evolutionary thinkers who emphasize the possible costs of attempting social change focus on four kinds of costs, as noted in chapter 2: (1) the investment of effort and resources required both to bring about (or attempt) the transition and to maintain the new state; (2) unhappiness and (3) the loss of freedom involved in suppressing or overriding people's natural tendencies (or attempting to do so); and (4) phenotypic trade-offs that must accompany any such change that is actually achieved. How ought these components to be weighed against each other? This question raises general and obvious problems about how to evaluate diverse costs and benefits in populations made up of people who differ from each other—and are socially connected to each other—in many different ways. As we will see, our assumptions about how internal and external factors interact to produce phenotypes and behavior have important implications for how we understand these problems.

Before moving on to consider the various kinds of costs separately, it is worth recalling from chapter 1 some general aspects of the accounting that evolutionary psychologists tend to overlook. First, if we are to make a serious assessment of the cost of change (or of a particular change), it is important also to include its benefits (or the cost of continuing in our current course); the former means very little without the latter. Second, it is essential to make explicit the unequal ways in which various costs and benefits are distributed across the human population. Pinker expresses distrust of uncritical discussions of what is good for "us" that fail to consider how reforms that benefit some people will inevitably harm others, but he and E. O. Wilson both often make this very error, writing as if the costs of change are shared equally by everyone (as when Pinker argues that excessive striving for equality is likely to result in a "leveling down" that will deprive everyone of the fruits of the talents that are thereby sacrificed).

Are the benefits of social change worth the costs? It is very plain that how you answer this question will depend on who you are. Determining the overall cost–benefit structure therefore requires explicit consideration of trade-offs between the positive and negative outcomes of the available options for different people. These points seem painfully obvious to many critics of evolutionary psychologists' warnings about the possible costs of social change, but their neglect by evolutionary psychologists makes sense in light of assumptions linked with these thinkers' quasi-internalist attitude. Recall that Pinker, for example, believes that the internal factors responsible for human preferences and behavior are so powerful that the aims of "utopians" are ultimately unachievable. In this case, the benefits of realizing those aims need not be reckoned, for they will never arise; only the costs of the (doomed-to-fail) attempt need be counted. Similarly, the tendency to neglect the crucial differences between the cost–benefit functions belonging to different people is of a piece with evolutionary psychologists' general willingness to abstract away from individual variation, treating population averages as the most important measure.

These (often misleading) simplifications—the overlooking of benefits and the uncritical use of population averages—are also standard features of cost–benefit analysis. They result from the assumption that all goods can be represented on a single scale, usually monetary. (In evolutionary applications, the scale measures reproductive success.) In contexts of policy choice, where benefits are less obvious or less easy to monetize than costs, they are apt to be overlooked, and classical cost–benefit analysis systematically ignores distribution issues: since costs and benefits are all counted in common units, they can be easily treated as belonging to a single pool. The standard assumption is that inequitable distribution of costs and benefits can be corrected by transfers of benefits among individuals. Evolutionary psychology and classical cost–benefit analysis thus share certain problematic simplifying assumptions about populations and social change.

The Problem of Value: What Is Good for Us?

Those who seek social lessons from evolutionary psychology have issued sharp warnings about paths of social change that we should avoid, but although they express these warnings in terms of costs, they have not attempted to support their conclusions about the prospects for change by means of classical cost–benefit analysis—and with good reason. Their discussion to date is far too broad and abstract to permit an actual cost–benefit accounting, and in any case the limitations of this approach to making decisions about public policy are widely recognized (Ackerman and Heinzerling 2004; Banzhaf 2009; Buss and Yancer 1999), and more sensitive methods (though with their own limitations) are available. Nonetheless, a serious consideration of the expected good and bad results—the benefits and costs, broadly construed—of the options that confront us is essential to our ability to make good choices by any of these methods, and even at a broad and abstract level of analysis some significant problems can be discerned in evolutionary psychologists' treatment of costs and benefits, goods and harms. The catalogue of costs that evolutionary psychologists discuss does make it clear that they (rightly) recognize the importance of costs that cannot readily be treated in monetary terms, or on a single scale. Their treatment of the various kinds of costs that they recognize, however, also reveals consequential biases resulting from their "limited malleability" perspective on the interaction of internal and external causes.

Evolutionary psychologists focus their discussion of happiness on the effects of situations in which (as they see it) people are constrained to behave in ways that they would not naturally choose. Many writers discuss historical examples of social environments that they see as incompatible with the expression of human nature and therefore as reducing people's happiness. Favorite instances include Israeli kibbutzim (where mothers were frustrated by communal child-rearing practices that prevented them from spending extra time with their own children) and essays in socialism or communalism (where people were dissatisfied with conventions that barred private ownership of goods or sexually exclusive relationships). They do not discuss the frustrations and dissatisfactions associated with the social arrangements that they see as relatively natural. Some of these are experienced by many people who do not conform to the social norms that are dominant in today's societies—including gender "outliers" such as lesbians and gay men, transgendered people, and others who simply have interests, preferences, or skills that are not consonant with existing gender stereotypes, and including interracial families, conscientious objectors, and many others who fail to conform with other aspects of the dominant social order or (presumed) "human nature." Even in the most liberal societies today, many people are actively blocked from pursuing activities and life paths that they prefer, or they are socially excluded, bullied, or subjected to physical harm as a result of the "natural" sex role differentiation or intergroup intolerance of the societies they live in. In other cases—by far the largest number—people's happiness is limited by poverty resulting from "natural" social inequality. For vast numbers of poor people, many important possible choices about how to live and how to develop their human potential are entirely closed. In fact, in some instances, almost the entire population of even an affluent nation may be barred from possibilities that would greatly increase their happiness.

Evolutionary psychologists also do not consider any compensating sources of happiness that might be found in the "unnatural" environments that they discuss, or inequalities in the distribution of effects on happiness across the populations involved. Nor do they mention the costs to human happiness that are likely to result from unavoidable changes to our social and material environments owing to climate change and population growth if we adopt a laissez-faire attitude to social change. The point is not that the kibbutzim or communes are better on balance than mainstream modern societies but that any assessment of these—and of other novel proposals for social change—must indeed be made on balance, with both positive and negative aspects of all the options taken into account (Agassi 1989). The choices that evolutionary psychologists make about *which* effects on *whose* happiness to count as costs—and which to ignore—thus betray a set of hidden value assumptions, some of which are rooted in their conception of human nature. Unhappiness that results when people are unable to act on preferences consonant with what evolutionary psychologists think of as human nature appears as a cost to be counted, but equal unhappiness resulting from the frustration of "unnatural" preferences does not. One example makes this contrast very stark. Several evolutionary thinkers write with feeling about an individual case that has become well known: a baby boy who was raised as a girl after a botched circumcision (Colapinto 2000; Diamond and Sigmundson 1997). As a child and youth, this boy was deeply unhappy and unable to reconcile himself with a female identity, and his troubled and ultimately tragic life is put forward as an object lesson that teaches about the limits that human nature places on our attempts to shape people's behavior by means of changes to their material and social environments. But the authors who draw this lesson entirely ignore the very similar and equally destructive suffering and confusion that many children growing up with gender identities that do not match their physiological sex experience as a result of the powerful gender socialization that we all undergo as a normal part of our existing social arrangements. Likewise, unhappiness that results from "unnatural" social arrangements is treated as a cost to be counted, but unhappiness that results from human nature taking its course is disregarded. The harm that would be done by "leveling down" is recognized as very serious, but the harm of existing social and economic inequalities is acknowledged much less clearly.

Evolutionary psychologists' treatment of freedom closely mirrors their treatment of happiness. They focus their attention on abridgments to people's freedom that have resulted or might result from particular social changes but omit to consider the many ways in which people living in today's societies already suffer from severe limits to their freedom. The problem of distribution again goes unaddressed, despite the special importance that it carries in this case, since some people may use their freedom to make choices that limit the freedom of others.

The intractable problem of how to count up the many interlinked goods and harms of social choices becomes more difficult still when we consider how comprehensively uncertain are the outcomes of such choices at the level of detail required for serious evaluation. Those who look to cost–benefit methods—even to more subtle variants on such methods—to make choices about broad issues of social policy may seem to be yearning for a determinacy we cannot expect to achieve. We often have to act in hope—being fairly sure that some courses of action are more likely to produce success than others but not knowing whether any have serious chances of success. That is what life is like. Could any form of evolutionary psychology so transform our cognitive predicament that we could plan with confidence? So far, the prospects seem doubtful.

It may seem that this chapter has dodged—or given up on—the really hard question: "How should people decide what is valuable?" In the next chapter, we consider a framework for approaching that question in a profitable way, one that will replace the inadequacies of cost–benefit analysis and help us to consider the full range of outcomes, good and bad, that are likely to follow from our choices.

[7]
Choosing Environments

EVOLUTIONARY PSYCHOLOGISTS' ACCOUNTING of the costs of social change in the form of happiness and freedom is, we have seen, both incomplete and biased. But there is a deeper problem. The very conceptions of happiness and freedom in their discussions of the costs of change are narrow ones that fail to take account of many important aspects of human well-being and autonomy (Kelman 2005).

The obvious place to start in assessing people's happiness in a particular society is to ask them how happy they are. But this approach has well-known weaknesses. Critics of welfare economics have noted the problem of "adaptive preferences"—that people modify their preferences and their standards of happiness to accord with their expectations (Gasper 2005; Giovanola 2005; Teschl and Comim 2005). Thus, people with lives that seem to be severely deprived by any objective standard may report that they are fairly happy and that their preferences are being met quite well simply because they have relinquished desires that they have come to regard as unrealistic. (People whose lives are exceptionally well provided, on the other hand, may be frustrated by their inability to fulfill overblown expectations and so report themselves to be unhappy.) Describing the phenomenon this way suggests that people whose lives are objectively deprived know more or less what they are missing though they have given up on it. But sometimes things are even more difficult for those seeking a happiness tally, and people's life circumstances are such that they really have no conception of what other, better lives might be possible for them.

One response to the problem of adaptive preferences is to replace the measures of subjective happiness or preference satisfaction with objective measures of people's ability to do things. This "capability approach," developed in different forms by economist-philosopher Amartya Sen and philosopher Martha Nussbaum, provides a way of assessing the quality of people's lives even when their own ability to judge their life quality has been compromised (Alkire 2002; Comim 2005; Gasper 2005; Giovanola 2005; Nussbaum 1995; Sen and Nussbaum 1993). The response perspective on development and behavior adds a further twist to the puzzle, however. It suggests that we need to consider not just the way that people may "downgrade" their preferences quantitatively in keeping with diminished expectations but also the likelihood that the different environments people inhabit may elicit qualitative variation in the desires that they come to possess. People sometimes want less because they expect less, but they may also want *differently* because they live differently.

Such differences could themselves be part of an evolved responsive human nature. When Sen talks about "adaptive preferences," he means that shifting preferences help people to adapt their attitudes to their circumstances, but shifting preference may be adaptive in another—evolutionary—sense as well. Some environmentally triggered differences in people's preference profiles may be evolved adaptations to particular kinds of environments. For example, Gowaty's "sexual strategies" hypothesis suggests that environmental factors may affect the extent to which both men and women develop behavioral tendencies and preferences that match common gender stereotypes (Gowaty 2003, 2008, 2011; Gowaty et al. 2007; Gowaty and Hubbell 2013). Just how strongly both women and men want to spend time with their children (or to pursue careers in the sciences) may depend on subtle environmental indicators of women's reproductive autonomy. Indeed, how strongly women want reproductive autonomy itself may depend on the indicators of that autonomy.

Other instances are equally suggestive. New thinking about the adaptive role of "risky behaviors" suggests that environmental factors may cue the adoption of life-cycle strategies that affect people's preferences regarding risk taking of many kinds (Chisholm 1999; Ellis et al. 2012; Wilson and Daly 1997), while recent work on the adaptive significance of bullying suggests that the kinds of peer relationships that people seek may be cued by background features of their social environment (Volk et al. 2012). It is worth looking at these instances a little more closely.

Risky behaviors are those that are likely to lead to harm for those who engage in them—typical examples include drinking and drug use, sexual promiscuity, petty crime and gang membership, and the like. It is well known that some social environments are much more conducive to risky behaviors among young people than are others. The conventional explanation of this pattern, the "developmental psychopathology" model, is a conservative interactionist explanation. It assumes that the normal developmental trajectory is one that leads via a well-adjusted childhood and adolescence to a responsible adulthood but that pathologies such as antisocial behavior or premature sexual activity can result when conditions in the family and local social environment do not meet the conditions required for normal development. The evolutionary explanation, instead, sees children as responding to their environments by "choosing" one of two available developmental strategies (although no conscious choice, of course, is presumed to be involved). If the environment provides cues indicating good prospects for a long and healthy life, the child develops according to a strategy that can take advantage of those conditions—deferring reproduction in order to achieve healthy maturity, high social status in adulthood, and advantageous mating, thus optimizing the prospects for successfully rearing high-quality offspring. But if environmental cues indicate that the prospects for a long and healthy life are poor, the child develops instead according to a "live fast, die young" strategy evolved to promote reproductive success despite environmental challenges—pursuing early mating and high social dominance in adolescence irrespective of the negative longer-term effects that are likely to follow (Amin and Thompson 2001; Chisholm 1999; Ellis et al. 2012). What kinds of cues are relevant? Research on the age at which women first become pregnant shows a wide range of factors. Girls who were separated from their mothers for long periods in infancy, whose fathers are absent, or who moved from one home to another repeatedly in early childhood tend both to mature earlier and to become pregnant earlier than their peers do (Nettle, Coall and Dickins 2011). Similarly, the conventional explanation for bullying understands it either as a pathological result of a lack of social skills brought about by a social environment inadequate to the needs of normal development or as learned behavior impressed upon developing children by bullying parents. The evolutionary explanation sees bullying instead as an adaptive response to particular cues in the social environment: a means of achieving social dominance in adolescence under conditions that make this strategy likely to succeed. Conventional explanations in both these cases see some desires—or some kinds of happiness—as proper expressions of human nature, others as distortions produced by substandard environments. The desire to find happiness by getting married and raising children in a safe neighborhood is part of a proper developmental path; the desire to find happiness by using illegal drugs and joyriding is a distortion. Evolutionary explanations, by contrast, see both desires as equally adaptive responses to different environmental challenges.

Environments thus seem to play two different but entangled roles in determining people's happiness: they help determine how capable people are of achieving the fulfillment of their desires, but they also help determine what desires people have. Is happiness a matter of the satisfaction of one's actual desires or of satisfaction of the *right* desires? Aren't some desires better than others, more likely to lead to true flourishing if they are fulfilled?

Sen and Nussbaum disagree somewhat about this—Sen considers that what matters most is the extent to which one has the capabilities needed to realize one's own preferences, whatever these might be, while Nussbaum places more emphasis on the idea that certain capabilities are critical for people to live good lives whether or not the people themselves recognize this. But the possibility of divergent developmental strategies reveals a different problem—that there are trade-offs between capabilities. Thus, if environmental differences can cue divergent developmental or behavioral strategies where each strategy leads to a different suite of preferences and capabilities, there may be no grounds for deciding which strategy leads to the better—or even just the happier—life.

The differences between these problems can be understood through the lens of dystopian—and utopian—fiction. Start by thinking about George Orwell's *1984* and Aldous Huxley's *Brave New World* (Huxley 1932; Orwell 1949). Orwell envisages a world of overwhelming oppression by a totalitarian government, one that prevents people from fulfilling their natural preferences, and seeks—with partial success—to eradicate the preferences themselves by brainwashing and torture. Huxley's vision is different—in his world, the government is not oppressive but manipulative; it does not block or suppress people's preferences—it shapes them. Huxley's World State uses people's natural responses to social agreement, bodily pleasure, and subtle (and not-so-subtle) environmental cues as tools to control their desires.

Evolutionary psychologists focus on the Orwellian threat of overt unhappiness resulting from people's inability to fulfill their preferences, or from forcible attempts to suppress those preferences, but in doing so they overlook a cost that is much more difficult to assess but potentially very significant—the Huxleyan danger of a hidden or cryptic loss of happiness resulting from a limitation of people's horizons that makes it impossible for them to form certain preferences in the first place. These environmental effects may be difficult to detect even for observers who are looking for them. They include instances such as girls who forgo mathematical training—and the intrinsic satisfactions, extrinsic rewards, and cognitive enrichment that it can bring—because stereotypes lead them to become convinced that they dislike math or are unable to learn it, men who similarly forgo the rewards of becoming primary caregivers for their children, and many others whose horizons are limited by stereotypes or socially enforced assumptions of which they are unaware. In some instances environmental factors may provoke developmental responses that affect the capacities and preferences that individuals come to possess at a neurological level. If Gowaty's hypothesis is correct, for example, environmental factors indicative of limited female sexual autonomy may trigger developmental responses that cause girls and boys to grow up with exaggeratedly different cognitive capacities and preferences, foreclosing permanently on some possible life paths for members of both sexes.

The further implication of the role of response in development, however, goes beyond both the Orwellian and the Huxleyan worries to what might be called a Le Guinean conclusion. Ursula K. Le Guin's novel *The Dispossessed: An Ambiguous Utopia* (Le Guin 1974; see also Davis and Stillman 2005) compares three fictional societies. Her focus is on the ambiguous utopia of the subtitle, a society based on a philosophy of cooperative anarchism distantly inspired by Jean-Jacques Rousseau. But this utopia is indeed ambiguous and flawed, and the alternative societies (representing the free market and collectivist options) are shown as having real (and differing) strengths. Le Guin's societies differ not only in the satisfactions that they provide but also in the desires that they foster, and because some of these are incompatible, no society enables its members to "have it all."

Le Guin's lesson is the lesson of the response perspective. Because different developmental or behavioral strategies will elicit different constellations of desires and capacities, there is no single form of the good human life, sanctioned by human nature; instead there are many different ways that human lives can be differently more—or less—good. As an abstract point, this seems reasonable—perhaps obvious. But it poses a difficult problem for practical decision making. How are we to choose between such alternatives? Suppose a simple intervention in an individual's environment could make the difference between two developmental pathways leading to lives with quite different preferences and satisfactions. Perhaps one pathway leads her to a gender-traditional life in which she develops a strong desire to have children with whom she has a close nurturing relationship, and this desire is fulfilled. Perhaps the alternative pathway leads to a life in which she has less interest in children but a strong desire for professional success and creative satisfaction as a musician, and this desire is fulfilled. Each life will also involve many other desires and many satisfactions and frustrations, large and small. If we had the power, early in a girl's life, to tip her toward one or the other of these pathways, how could we tell if it is right to intervene? This problem is hardly resolved by pointing to population averages or "species-typical" outcomes as indications of "human nature."

Similar considerations apply, once again, to discussions of freedom. Evolutionary psychologists focus on a narrow conception of freedom as the ability to act in accordance with one's preferences, and they regard people's preferences as fixed relatively firmly by evolved human nature. They thus overlook the possibility of "adaptive preferences" both in the economist's sense (preferences adjusted to avoid intolerable conflict with people's expectations) and in the more radical sense (preferences that vary as part of environmentally cued variation in evolved developmental and behavioral strategies).

But their simple conception of freedom, and of its relationship to happiness, has deeper problems than this. The facile presumption that when people—members of a population—can choose freely, they will maximize their happiness, for example, is a long-recognized and serious error. Where people are free to express and act on their prejudices, one of the things they do is to limit other peoples' choices. People's freedom is restricted when they are barred from some domains of activity by their own and others' responses (such as fear of social exclusion or biased hiring practices) or by niche construction (such as social and physical environments that exclude some kinds of people). And people's happiness is reduced where they make their "free" choices within environmental contexts that attach intolerable social costs to options they would otherwise have preferred. When girls are bullied for expressing an interest in stereotypically "male" science fiction or computer gaming, boys are bullied for showing any enjoyment of "female" play with dolls, or black students experience social rejection for showing an interest in "white" academic success, their choices cannot ensure their happiness.

The capability approach gives a broader conception of freedom not merely as freedom from constraint but as freedom to accomplish things—to act autonomously, to realize one's potential. According to this richer conception of freedom, very many people who are now able to follow their preferences relatively unimpeded are nonetheless far from free. And impending changes to our global environment will surely further drastically diminish many people's freedom—taken in this broader sense—unless we are able to respond to these changes with appropriate social reform. Yet as in the case of happiness, a puzzle remains about how to compare lives with different constellations of preferences and capabilities, lives that realize different and incompatible sets of potentials.

The problem of how to weigh the trade-offs between alternative possible developmental trajectories for an individual and the lives these might lead to is already a difficult one. But it is complicated further when we consider change at the social level—interventions capable of affecting the lives of many different individuals in different ways at the same time. Thinking about capabilities as a key measure of a good human life can help to clarify what is going on in discussions about phenotypic trade-offs in this broader context.

Consider some characteristics that are good for people to have—phenotypic traits that we see as contributing, directly or indirectly, to people's flourishing: good health, intelligence, compassion, curiosity, linguistic facility, physical coordination, self-confidence, musical ability, social sensitivity, mathematical aptitude, a sense of humor, and so on. Thinking about "happiness" and "freedom" is too abstract—it is important to consider the specific characteristics in which these abstractions are realized in order to take account of the different ways in which they contribute to good human lives. According to the capability approach, what is of value is not these phenotypic characteristics themselves but the contribution that they make in particular environments to people's capabilities and resulting experiences—e.g., to achieving mastery in particular areas of human endeavor such as mathematics, language, sport, music, or the arts; to acquiring rich knowledge of the natural world or of cultural traditions; to exercising social leadership, making good judgments, showing compassion, sharing aesthetic experiences, creating works of art or novel scientific hypotheses, or building close social relationships.

But the phenotypic characteristics are crucial nonetheless, for they provide the basis for these capabilities.

How can we compare the value of these various traits and the capabilities they support? As we noted at the outset of this book, there are problems both about how to weigh different phenotypic goods against each other and about how to weigh individual outcomes against distribution across the population. Consider again a situation in which we have the chance to make an intervention, but this time imagine an intervention capable of changing the phenotypes of a whole population, increasing or decreasing individuals' performance on measures of some traits. Some such interventions can change the range of variation itself, modifying the peak performance—the highest performance achieved by any members of the population. We have three options:

- *Option A* increases peak performance on some measures but decreases other measures for the same individuals.
- *Option B* increases peak performance on some measures but decreases peak performance on other measures affecting other individuals.
- *Option C* decreases peak performance on one measure but increases performance on the same measure for many individuals at lower levels, increasing equality on that measure across the population.

There is no way to compare these outcomes to each other or to the status quo in the abstract; if the question is which option is the best choice, different answers will be appropriate for different measured traits depending on the kinds of value that they supply. In many cases it seems likely that even quite rich information about the effects of a particular intervention would leave its cost–benefit ratio hard to determine simply because the goods involved are so disparate. This situation makes such assessments particularly susceptible to bias resulting from the tacit assignment of relative values or weights to particular goods. The examples favored by evolutionary psychologists suggest that they are inclined to overvalue goods that they deem "natural" and to undervalue those they deem "unnatural." This bias is widely shared, of course, as is a general tendency to assign high value to goods that are accorded high status in our culture. Both of these biases can contribute to assessments of the costs and benefits of change that echo existing gender stereotypes.

Option C on the list raises a special problem: how to weigh the value of equality against the value of maximizing peak performance on the variables that are deemed desirable. This is an important type of case for evolutionary psychologists since one of their main claims about the prospects for social change is that equality between the sexes, on many measures, is achievable only at a cost that might well be unacceptable. Recall Steven Pinker's argument in favor of providing the special environments needed for rare talents to be realized and against the "leveling down" that a simple-minded egalitarianism might lead to. Pinker notes that we all benefit from the realization of rare talents, and so we all suffer if those talents go unfulfilled as a result of the pursuit of equality. But this is too quick, for we might all also benefit from the full realization of the unexceptional capacities of large numbers of people as well as from equality itself. The question is, how much? No doubt there are some traits for which what really matters most is what happens at the peak—for which a few extraordinary performers make a large difference for everybody, or at least for many. Pinker is thinking of cases like that. But a lot of work would need to be done to establish which traits really work like this. Even in the cases that appear to be of this sort, the assessment seems quite speculative. Do we really know that a single Mozart or Einstein is worth more than thousands or millions of more modest achievers in the arts or the sciences? And on the other hand, just how should we assess the loss to the most talented, if they are unable to develop their talents? These questions are not merely hypothetical. Finland, for example, does exceptionally well at educating a wide range of children. Finns sometimes complain, however, that their educational system does not fully develop the talents of the most gifted.

The problem of how to compare different distributions of goods is complicated by the possibility that equality or justice may have value of its own, as many philosophers have supposed. (It may also have very broad beneficial social effects, as some social scientists argue [Wilkinson and Pickett 2010], but this is a different point.) The "utopian" proposals for social change that evolutionary psychologists call into question often do take social equality as a serious goal. As I note in chapter 1, this is a goal that evolutionary psychologists often dismiss out of hand on the assumption that in many cases equality can be achieved only by means of a destructive "leveling down." This assumption concerns issues of feasibility, and some reasons to reconsider it are offered in the next chapter. For now, it is sufficient to note that the very important and separate question of what the value of equality itself might be cannot even be raised within the individualist frameworks of classical cost–benefit analysis or evolutionary psychology.

A related point concerns the kind of equality in question. Social reformers often take equality as one of their highest goals, but as Amartya Sen has asked (Sen 1992), equality of what? Evolutionary thinkers discuss equality in two main contexts: in arguing for the impossibility of achieving "perfect gender equality" in behavior and in considering the harmful effects of "leveling down" in pursuit of equality. In both cases their focus is on (exact) equality in phenotypic outcomes. This is the kind of equality Kurt Vonnegut describes in "Harrison Bergeron":

The year was 2081, and everybody was finally equal. They weren't only equal before God and the law. They were equal every which way. Nobody was smarter than anybody else. Nobody was better looking than anybody else. Nobody was stronger or quicker than anybody else. All this equality was due to the 211th, 212th, and 213th Amendments to the Constitution, and to the unceasing vigilance of agents of the United States Handicapper General. (Vonnegut 1961, 5)

But though Vonnegut's story is read by almost everyone (including Pinker) as a short, crude analog of Orwell's *1984* or perhaps of Ayn Rand's *Anthem*, its point is something quite different. It is a satire not of egalitarianism but of the fear of egalitarianism, and one of its main targets is the very idea that equality "every which way"—simple equality in people's qualities and abilities—is what egalitarians aim to bring about (Hattenhauer 1998). The question of just what form and degree of equality is a good social goal is a difficult one. Sen offers a promising approach in *Inequality Reexamined* (Sen 1992) based on the capability view of what makes a good life. He argues that the inequalities that are harmful are inequalities in people's abilities to achieve the ends that they value. This view is explicitly intended to make sense of how we can strive to reduce harmful inequality while recognizing that people differ from one another in many important ways so that both their preferences and their responses to particular features of their environment are diverse.

The costs of change that we have considered so far have taken the form of side effects: unintended and undesirable effects of actions undertaken to achieve desired aims. Costs of another sort, what I have dubbed investment costs, appear to fit into the story in quite a different way: as expenditures of resources or effort deliberately undertaken in order to bring about the social change in question or to maintain it once achieved: "the added energy for education and reinforcement," as E. O. Wilson put it (1978, 148).

From the perspective of ordinary cost–benefit analysis, this conception of cost makes eminent sense; indeed, it is the only kind of cost that classical cost–benefit analysis—which weighs expenditures against payoffs—normally counts. Yet in the present context there is something peculiar about this notion of "cost." Social change undoubtedly requires that resources and human effort be spent to achieve the transition, and perhaps to maintain the new state of affairs, and this expenditure is usually counted as part of the cost of such change. But tremendous quantities of resources and energy—in the form of education, advertising, cultural production, legislation and law enforcement, and individual social activity ranging from gossip and peer pressure to threats and violence—also go into reinforcing the patterns that evolutionary psychologists term "natural." Should these be counted as costs associated with the status quo?

Many optimists about human possibilities think the answer is yes. These are investments of resources and effort that people are choosing to make—expenditures by individuals, by businesses, and by public institutions. We could make other choices instead without changing the amount of the investment. If all the resources—human and material—that we now spend on communications and activities that reinforce gender and racial stereotypes, encourage selfish consumerism, and aggrandize violent aggression were spent instead on alternatives that encourage generosity, compassion, and nonstereotypical thinking (optimists suppose), we might achieve substantial changes in the human condition with no extra investment cost at all.

This hypothetical possibility raises obvious questions about freedom of choice and about happiness in the form of people's ability to live according to their preferences. The current allocation of resources, including human effort, is often presented as being the result of free choices in a free market. Advertisers use stereotypes to sell their products (it is said) because customers respond to them; movies and video games offer endless representations of men in violent conflict and women as sexual objects because that is what viewers want. This perspective sees the current use of resources not as a cost or expenditure but as an expression of various aspects of human nature. From this perspective, any sort of requirement that those resources be diverted to sending other messages looks like a combination of censorship and propaganda—it would obstruct people's freedom to choose what messages they send, what cultural expressions they attend to, and what activities they pursue.

Philosophers have long debated whether existing social institutions *express* human nature or *restrict* it. Thomas Hobbes saw social institutions as the natural outgrowth of human nature as well as the only possible means of checking its excesses; Rousseau saw many social institutions as the chains that prevent human nature from expressing itself fully. The niche construction perspective makes it clear that this question needs rethinking: institutions (and many other human activities as well) have a double face—they express our nature but they also shape its expression. The choices that people make (as legislators, marketers, teachers, artists, consumers, parents, or community members) limit the choices available or even conceivable to others and to themselves. It cannot simply be claimed that the expenditures that people freely choose are not costs but an expression of their nature, whereas expenditures imposed upon them must be counted as costs. People make different choices under different circumstances, and their very choices constantly impose changes on others' circumstances (and their own) that in turn shape the further choices of those affected. Indeed, we are evolved to struggle for control over ourselves, our environments, and each other. No choices are "free" in the sense of being independent of the choices others have made. The distinction between "freely chosen" and "imposed" expenditures of effort and resources cannot ground a distinction between expenditures that do not count as costs and those that do.

Is there another way to distinguish costs from other human expenditures of effort and resources? One intuitive idea is that some such expenditures are themselves natural to us and so do not count as costs. But this is again to accept an internalist distinction between phenotypes that are "natural" and those that are "forced" by environmental circumstances. A more promising way forward might be not to distinguish between natural and unnatural phenotypes or natural and unnatural environments but to tackle both together in the form of constructed niches and the traits that shape and respond to them. Perhaps some social systems—some combinations of social and material environments and associated phenotypes—are special in that they are self-sustaining, whereas others require ongoing intervention to keep them going. This idea is suggested by Wilson's mention of the cost of "reinforcement" of new behavior patterns. It is made more explicit by discussions by Pinker and Buss of attempts to create new social systems that failed because the combinations were not self-sustaining. The fate of the utopian communities of the late nineteenth century and communes of the late twentieth century and of the more radical features of the kibbutzim, according to these evolutionary psychologists, reveals how human nature destabilizes social systems that are unnatural for us. No extra push is needed to keep people doing what is natural and recreating the environments that in turn support them in doing so, but the unnatural requires ongoing investment to block reversion to the natural state (and is likely to fail, even so). Natural social systems, according to this perspective, are thus those that are self-sustaining; the use of resources by which they sustain themselves should not be counted as a cost but as part of their functioning.

For this view to make sense, the notion of "self-sustaining" systems—as distinct from those that are sustained by paying a cost—must be made clear. One obvious sense in which a system can fail to be self-sustaining is if its maintenance depends on inputs from an external source. This is the sense in which "natural" ecosystems and some traditional forms of agriculture are self-sustaining while modern agriculture that depends on inputs of fuel and inorganic fertilizers is not. Some local social systems certainly fail to be self-sustaining in this sense in that they are maintained only by reinforcement from without: prisons and colonial social systems are obvious examples. At the level of whole societies, however, this way of thinking seems to break down; the means by which fragile features are shored up (such as government funding or law enforcement) are themselves internal to the society. At this level of organization we can see more clearly that what matters is the robustness of the system as a whole: its capacity to reproduce and reinforce its own structural features, including the mechanisms by which its most fragile elements are maintained, and so both to survive external changes and internal disruptions and to renew its capacity to keep so doing.

The social systems that evolutionary psychologists regard as "natural" may well be robust in this sense, but it does not follow from this that other—and quite different—robust social systems are not possible. Whether such alternatives exist is a key question for any attempt to gauge the feasibility of social change and is addressed in the next chapter. An equally important question concerning the value implications of robustness has already been answered. There is no question that some very robust social systems are not good for human flourishing—a look at history makes this clear. Robustness is not itself a good and is no guarantee of goodness—if a social system makes people miserable or diminishes their lives, the fact that it is robust is no compensation; quite the contrary.

What I am suggesting is that the response perspective demands a shift not just in what counts as a cost but in what counts as an act or environmental condition whose costs must be reckoned. We cannot set aside some acts and conditions as "natural" and therefore to be omitted from our accounting; they must all be counted. The question is not whether they are natural but how robust are the social systems that give rise to them and what effects do they have on human lives. The measure of investment cost disappears, therefore, to be replaced with a measure of the negative effects the proposed new social system will have on human flourishing and a measure of how difficult it will be to achieve and maintain. This last is not a value measure at all but something equally important—a measure of feasibility.

[8]

What Is Feasible?

TO MAKE GOOD JUDGMENTS about what sorts of social change to pursue, we need to assess the comparative value of different social arrangements, but we need also to gauge their feasibility. When evolutionary psychologists criticize utopian thinking, one of their main points is simply that many social arrangements that are perfectly conceivable and highly desirable cannot be realized by human beings—and of course this is true. But like assessments of value, judgments about the feasibility of social change turn out to be more complex than they might at first seem, and they depend crucially on how we think about the interaction of internal and external factors in driving human development and behavior.

For a particular social change to be feasible, it must be both achievable and sustainable—we must be able both to "get there from here" and to stay there (for long enough to be worthwhile) once we have arrived. Both of these dimensions seem plainly to permit differences of degree: some changes that are possible would nonetheless be very difficult to achieve or sustain while others might be easy. The prospects for both achievability and sustainability of social change look very different as seen from the response perspective than they do from the broadly internalist perspective of conservative interaction (or "limited malleability").

As evolutionary psychologists see it, stability of social arrangements is either an expression of human nature or something precariously sustained by pushing against human nature. The status quo in liberal democracies is presumed to be something close to the natural state of complex human society so that sustaining it requires no special effort. To achieve social change of certain kinds, however, requires breaking away from this natural state, and sustaining such change (it is supposed) requires constant resistance to people's natural tendency to backslide. Both of these require powerful "forces." Seen from the response perspective, on the other hand, stability is a matter of robust social order actively reconstructed by a richly interconnected web of responses. The status quo is maintained this way, so achieving change may be as easy as disrupting one or more of the links by which the existing social order reproduces itself or making an intervention that creates new links, thus triggering a shift to a new robust state. This may be surprisingly easy in some cases but by no means in all. Some interventions may hit effective leverage points while others, intuitively equally promising (or more so), may be remarkably ineffective. Determining which interventions are effective will be a major task—in particular, finding the unrecognized links now reinforcing the status quo and the unexpected leverage points capable of producing desired changes.

These ideas are very abstract, but recent research in cognitive psychology, social psychology, economics, and neuroscience offers insight into the nature of some of the kinds of causal factors involved in sustaining current social arrangements and the kinds of interventions that might shift them. This research is progressing rapidly; what follows offers only a glimpse into the fascinating possibilities that it is beginning to uncover.

Some features of the current social order are sustained by patterns in individual choices, and some of these patterns can be modified by adjustments to the choosers' environment—what economists call "nudges" (Sunstein 2014; Thaler and Sunstein 2008). Thus, for example, children tend to choose unhealthy snack foods in school cafeterias, but this pattern can be modified by putting the candy bars and potato chips in an out-of-the-way location and displaying healthier foods in places that are easy to see and reach. Workers usually tend not to direct enough money to their retirement plans, but if their optional contributions are reconfigured as "opt-out" rather than "opt-in" choices, their tendency to stick with the default option works to increase rather than decrease their savings. Another example of the difference between opting in and opting out is that of organ transplants: many European countries take the default to be that organs of people killed in car accidents can be used for transplantation, whereas the United States has an "opt-in" approach. Unsurprisingly, in the countries that require motorists to opt out, many more organs are available.

These examples seem to accord with common sense, but social psychologists have found many similar effects that may be more surprising. Research shows, for example, that people who are wealthy—or who feel wealthy after playing in a rigged Monopoly game—are more likely to behave selfishly or unethically in social interactions (Piff 2014). But this effect can be defused by a small local intervention that provokes a feeling of compassion, such as watching a short video about children in need. Other studies show that people are more likely to be helpful, generous, compassionate, or trusting if they have recently spent time in a natural setting or even looked at a beautiful plant or an image of beautiful natural scenery (Zhang et al. 2014; Piff et al. 2015); people also tend to discount future costs and benefits less when they make decisions in natural settings (van de Wal et al. 2013). These effects are usually local, and the environmental adjustments must be maintained for their effects to continue, but some of them have the promise to enable substantial social changes simply by modifying the local environments in which people make the choices that shape the paths of their lives and those of others. Note that some of these are changes that have to be actively recreated over time while others could be maintained by building them into the physical or social structure of our communities.

Many features of our current social world are sustained by how people think about things—by the identities, evaluations, and narratives that we use to interpret our experience. Some of these cognitive patterns turn out to offer remarkably powerful leverage points for intervention. At their best, these can enable what social psychologist Gregory Walton has called "wise interventions"—interventions grounded in psychological theory, finely tuned to act upon a particular psychological process in a way that has powerful downstream effects, and aimed at enhancing human flourishing (Walton 2014).

Among the most striking wise interventions are those involving stereotypes. Stereotypes are one form of what cognitive psychologists call "schemas": simplified conceptual representations of people, things, and processes that allow us to negotiate the complex and chaotic world we live in. Stereotypes are schemas representing categories of people: women and men, boys and girls, blacks and whites, rich and poor people, scientists and executives, environmentalists and politicians. Like all schemas, stereotypes can be very useful (indeed, essential) to us, simplifying the massive social complexity we face enough to make it manageable. But like other schemas, they can also be misleading and—if they are mistaken—difficult to correct. Stereotypes make the people who hold them prone to confirmation bias—the tendency to overestimate the importance of supporting evidence and to underestimate the importance of counterexamples (Nelson 2014). This can make self-correction practically impossible.

Stereotypes are created and sustained by social processes: by people repeating and passing on judgments and stories and—

overwhelmingly, today—by the news, entertainment, and advertising media. New stereotypes often build on those that already exist, but they can sometimes develop impressive new strength rapidly. The stereotype of young girls as "little princesses," for example, was based on longstanding elements from European folk culture but was aggressively promoted by the media franchise Disney Princess through movies, television, and merchandise advertising and had become remarkably pervasive in North American culture within a decade of the franchise's launch in the late 1990s. Stereotypes reinforce existing social patterns by shaping people's behavior toward one another as well as their sense of their own social identity. In a society with strong sex-role stereotypes, girls and boys are exposed to very different social environments from birth, for example, and develop very different expectations for their own lives. Stereotypes can also reinforce other social patterns such as social hierarchy, interracial tensions, and the suppression of sexual diversity. Some stereotypes are based on "reality"—that is, on independently existing patterns—while others are not, but even those that are can reinforce social patterns that would otherwise be less marked. The stereotype of older people as slow moving is undoubtedly based on an independent reality. Nonetheless, that stereotype itself has effects that reinforce the existing pattern. It causes others to treat older people in a way that expresses the expectation that they will be slow and hesitant, and (when activated) it causes older people actually to move more slowly (Levy 2003).

Twenty years ago many psychologists regarded stereotypes, once acquired, as essentially immutable. Recently, however, a great deal of research effort has been dedicated to trying to find ways of counteracting or undermining harmful stereotypes. Several methods have shown promise, including giving people training in "media awareness" and critical thinking skills, exposing them to media representations that challenge or disconfirm common stereotypes, and providing direct social contact with members of stereotyped groups. This research suggests that the stereotypes held by individuals can indeed be changed, at least temporarily. Most experiments do not attempt to assess long-term effects, but a few are promising, showing that people's attitudes and behavior showed effects several months after the experimental intervention (Paluck 2009; Paluck and Green 2009).

One important effect of stereotypes has been discovered more recently: the phenomenon known as "stereotype threat" (Forbes et al. 2012; Keller 2007; Nosek, Banaji, and Greenwald 2002; Nosek et al. 2009; Steele 1997; Walton and Spencer 2009). You are likely to suffer from stereotype threat if you are a member of a group that is stereotyped as performing poorly at a particular type of task, you are asked to perform a task of that type, and you believe that your performance will be judged by others. Thus, white men asked to perform a physical coordination exercise that they are told is a test of natural athletic ability, or girls asked to take a math test, are vulnerable to stereotype threat. If stereotype threat is activated, it engages the threat response systems of the brain, increasing the person's heart rate, respiration rate, and stress hormone levels along with the level of activation of the "fight or flight" control region in the amygdala. Like other forms of threat arousal, it has the side effect of causing the person to perform poorly at many kinds of tasks—brain resources have been recruited for higher-priority self-protective functions and are not available for ongoing tasks.

Experimenters measure the effects of stereotype threat by finding a way to disarm it and comparing the same people's performance on the same task with and without stereotype threat. Stereotype threat can be disarmed in several different ways: by undermining the stereotype of poor performance upon which it is based, by priming the subjects to think of a different group identity (one not subject to a relevant stereotype), or by presenting the task not as a test but as a learning opportunity. Take the instance of girls asked to take a mathematics test. To undermine the stereotype, experimenters told the girls that the particular test they were taking was carefully designed to be equally difficult for boys and girls. To activate another group identity, Asian girls were asked to color in a drawing of a pair of chopsticks, prompting them to think of themselves as Asian (a group stereotypically thought to be good at math) rather than as girls. To present the task as a learning opportunity, experimenters simply described it that way rather than as a test. In all of these "disarmed" circumstances, girls performed much better on the test than they did under "normal" conditions; indeed in most cases the "gender gap" separating male and female performance on mathematics tests disappeared entirely (Steele 1997; Fine 2010). These experiments show that the effect of stereotype threat is relatively large: girls and women given math tests under stereotype threat conditions (e.g., asked to check a box specifying their sex immediately before taking the test) underperform those taking the same test under "disarmed" conditions by a full standard deviation.

Stereotypes and stereotype threat combine to form a destructive feedback system: the stereotypes drive the stereotype threat response, but that response in turn causes people to behave in a way that confirms and strengthens the stereotype both for those directly affected and for others. This feedback structure is what makes stereotype threat so destructive, but it also means that effective interventions can themselves trigger positive feedback cycles, leading to startlingly large and long-term results. Psychologists extending the research just described have found that interventions designed to invoke people's sense of themselves as strongly integrated individuals rather than as members of a group can provide long-lasting protection against stereotype threat. African American university students who wrote a short essay in class about their personal values and why those values should be important to others completed the semester with higher grades than control students; the effect continued for two years after the brief intervention and, indeed, was greatest at the last point measured (Steele 1997). This stereotype threat intervention is a good example of a "wise intervention"—small, precisely tuned, and with a large and lasting effect produced by positive feedback to enhance people's capabilities.

Stereotypes are not the only beliefs that can perpetuate existing social patterns. Carol Dweck has become well known for experiments showing the remarkable effects of two different attitudes toward intelligence and mental abilities, what she calls the "growth mindset" and the "fixed mindset" (Dweck 2000). People with a fixed mindset believe that one's intelligence is fixed (genetically or otherwise) and that mental abilities ("talents") are also essentially static—either you have them or you do not. Those with a growth mindset believe that intelligence and mental skills can be strengthened with practice ("like a muscle"). Dweck's research shows that converting underachieving students from a fixed to a growth mindset can have remarkable effects, allowing them to raise their level of performance while comparable control students continue to slide. Students with a fixed mindset are unwilling to risk failing at any task (since they think failure shows that they are inherently incapable) and unwilling to learn from failures when these occur. Students with a growth mindset, by contrast, are willing to take greater risks and to treat failures as learning opportunities. In a typical experiment, students learn the growth mindset by studying an essay about neural plasticity and learning, and then take possession of the ideas by explaining them to somebody else. Here, as in the interventions aimed at disarming stereotype threat, a pattern of recursion or positive feedback enables a single brief intervention to have effects that continue to build over time as the "growth" hypothesis becomes self-vindicating—the intervention tips an individual's process of psychological development from one pathway to another that diverges from the first increasingly over time.

Existing social patterns may also be reinforced by means of larger material and social circumstances that shape people's development and behavior. I have already mentioned how wealth affects the ways people feel and act toward one another. Poverty also has marked effects. Poor people are likely to be more generous and empathic than wealthier people (Piff et al. 2010, 2012; Varnum et al. 2015), but they also suffer from distraction and diminished ability to plan effectively and make good decisions as a result of the cognitive and emotional load caused by their financial insecurity (Mani et al. 2013). These patterns contribute to the deepening entrenchment of wealth inequality. Similarly, homelessness is strongly self-perpetuating since a person who is homeless will face serious obstacles to finding and keeping a good job. Giving, money or homes to poor individuals can free them from the disabling side effects of poverty and enable them to employ their own capacities more effectively (Hanlon et al. 2010). When effective, such interventions produce self-sustaining change by means of positive feedback.

Other compelling examples also show the potential for larger-scale social or material interventions in people's environments to trigger recursive change in their behavior in such a way as to disrupt entrenched social patterns. Students from disadvantaged backgrounds who win entrance to elite colleges tend to perform far below the levels that would be predicted for them based on their test scores alone (Steele 1997). Stereotype threat is partly to blame for this pattern, and interventions to disarm it are quite effective in closing the performance gap. But other factors are involved as well. These students are often socially isolated in college, lacking the sort of social network that provides both emotional and practical support for other students. An intervention intended to provide such students with an effective social support network has grown into the Posse Foundation, an organization that places teams ("posses") of students from nontraditional backgrounds at elite colleges and universities, providing each team with support and training to enable its members to help each other through their degrees. This remarkably effective high-investment intervention produces long-term, self-perpetuating change in young people's lives (Bial and Rodriguez 2007).

A final example is also in active use in educational institutions. The Roots of Empathy Project changes existing patterns of aggression and bullying among schoolchildren by means of a social intervention: a parent brings a baby to class once every few weeks, encouraging the children to watch the baby and talk about what it is feeling. The children receive related instruction about infant development and coaching in how to label emotional states. The effects of this intervention are durable. In one large study, improvements in levels of violence, bullying, aggression and prosocial behavior were found three years after the children had completed the program, and many measures were continuing to improve at that point (Santos et al. 2011).

These last examples begin to suggest a further possibility: interventions that produce changes that not only are self-perpetuating for the individuals directly affected but that also spread beyond them to affect a larger community in ways that might be self-reinforcing. The Posse Foundation program changes the lives not just of the students it enrolls but of their families and other members of their immediate communities, especially of their children. Similarly broad effects are likely to flow from the Roots of Empathy program. Some studies are beginning to explore the social "contagion" of behavioral changes triggered by social interventions. In an experiment on high school students, for example, a few students were given "antiprejudice" training and encouraged to intervene with their peers in response to discriminatory speech or behavior. Five months later, friends and acquaintances of the students who received the training were more likely to sign a petition supporting gay rights than control students (Paluck 2011).

Such instances provide a clear indication of the striking effects that relatively minor environmental changes can produce if they are correctly tuned to human sensitivities and patterns of response. The interventions they involve look remarkably unlike the heavy-handed methods that evolutionary psychologists had anticipated would be required to produce greater equality of intellectual performance or to decrease aggression and intergroup hostility, yet these are the effects achieved.

Steven Pinker's recent work engages more actively than most evolutionary psychology has done with the ways in which environmental adjustments can change human behavior and foster social progress, and develops a richer view of how social change can come about. In *The Better Angels of Our Nature: Why Violence Has Declined* (2011), he examines the aspects of evolved human nature that enable us to progress toward a less violent way of life. These are the "better angels" of human nature; they include empathy, self-control, the moral sense, and reason. They are opposed by another set of motivations, the "inner demons," which include the inclination to harm others if it furthers our own interests (predation); the desire for power and prestige (dominance); the urge for retribution (revenge); the enjoyment of others' suffering (sadism); and the commitment to shared belief systems that justify violence, often by appeal to utopian goals (ideology). Pinker argues, on the basis of an extensive statistical analysis, that violence has declined dramatically over the span of human history, and especially in the last few centuries. His statistical treatment of non-state societies is selective and may be misleading, but this does not undermine the most important point of the book. Pinker's approach is to explain the decline of violence by looking for the changing circumstances that have favored peace-conducive human motivations over violent ones, the angels of our nature over our inner demons. Whether his list is complete or correct, and whether the historical trends are as clear as he suggests, the approach is a good one. Some societies are certainly less violent than others. The way to explain this is to uncover the environmental features for those living in each society—including those created by the society itself—that elicit the particular combination of violent and pacific responses which are responsible for the society's overall degree of violence.

The circumstances that Pinker identifies as crucial to the decline of violence are the creation of the modern state and apparatus of law enforcement and justice (what Thomas Hobbes called the "Leviathan"); the rise of systems of trade that connect ever wider networks of people, making them dependent on each other for economic success and more apt to empathize with each other (commerce); an increasing public role for women, and growing respect for women's perspective (feminization); the development of communications media and patterns of travel and education that encourage people to learn about other cultures and perspectives (cosmopolitanism); and an increasing cultural emphasis on using reason to solve problems (reason). Pinker's explanatory analysis is illuminating and valuable; its main flaw is that it is too narrow. Pinker's focus is on a few key forms of social niche construction—large-scale changes in the social practices and beliefs of human societies—and continues to assign a key role to the use of force or the threat of force by the State. He neglects the other kinds of change—changes in the material environment, including tools and clothing and in fine points of people's bodily interaction, for example—that may act as triggers for cascading changes. And he underplays many of the deliberate ways in which people adjust their environments to modify each other's behavior and their own that work directly rather than via the creation of state authority and the judiciary. In addition, of course, his focus is on violence alone—a large enough topic, as he notes, but not the only one that matters. To think effectively about the possibilities for social progress we must consider a much broader range of possible environmental factors and corresponding behavioral and developmental responses.

The response perspective makes clear why some small interventions may have substantial effects on people's patterns of behavior while other and larger interventions may have none. By the same token, it indicates that some small environmental factors now operative may be playing roles in maintaining the status quo that are quite disproportionate to their apparent importance. Some apparently trivial differences in the experience of girls and boys—including some tiny injustices or what ethicists call micro-inequities—may lead to significant divergences in their development or behavior if they affect variables that serve as cues controlling individuals' "choice" of developmental or behavioral strategy. Some apparently trivial features of the physical and social niches we create may in turn have significant effects on the generation of micro-inequities. The same goes for other putative aspects of "human nature": microfeatures of the environments we create may play important roles in maintaining the patterns that appear to be "natural" for us (Friedman and Sutton 2013).

Of course feminists and other social critics have understood part of this point for a long time, pointing out that small factors that affect our self-conception or that have symbolical significance (such as what words we use to describe categories of people and what images of people we publish and display) may have far-reaching effects (Valian 1998). But the response perspective reveals further implications: that in addition to these meaning-based leverage points, there are likely to be others that are not intuitively plain and that may seem to have no symbolical or comprehensible significance at all, and some of these may affect people's development in ways that have lasting effects on their life prospects. We have barely begun to build an understanding of what the leverage points might be for reinforcing or modifying current patterns of human development and behavior, and what range of outcomes they can enable (Wilson et al. 2014).

The response perspective does recognize our beliefs about people—ourselves and others—as a key channel of influence through which many aspects of our development and behavior are cued. This has noteworthy implications for thinking about the costs and benefits of the choices we make. In particular, it poses a challenge to the distinction between "mere discussion" and the actions whose costs must be reckoned. When people in positions of authority make assertions about limits imposed by human nature, these statements may themselves constitute substantial and harmful interventions in the world: they affect others' implicit biases, experience of stereotype threat, and belief in the fixity of their own capacities, all of which contribute strongly to the suppression of some people's potential. Understanding this helps to makes sense of an important and difficult area of conflict between evolutionary psychologists and many of their critics. If, in the spirit of conservative interactionism or internalism, you think that only large environmental interventions can have any substantial effect on people's development or behavior, then the difference between "mere" speech and the "sticks and stones" of practical action will appear fairly categorical. But if, in the spirit of radical interactionism, you think that small interventions can sometimes have surprisingly large effects, speech becomes a potentially very important form of action.

Feminists and others have sometimes been sharply critical of evolutionary psychologists—and others persuaded by their work—who use their institutional authority to promulgate ideas about limits imposed by human nature. In perhaps the most notorious instance, women scientists walked out of the room in protest when then Harvard president Lawrence Summers asserted that we ought to entertain the hypothesis that the absence of women at the highest levels of achievement in mathematics and related fields is the result of "hardwired," evolved differences between male and female brains. When events like this take place, evolutionary psychologists and their allies point to the value of free speech and open debate—to "what makes a university different from a madrassa," as Pinker says (2005). Whether the hypothesis that Summers pointed to is true or not, they say, it is crucial to our ability to learn the truth and act upon it that we are able to entertain and discuss that hypothesis and others we may like even less.

But critics see statements like Summers's as consequential acts as well as contributions to a discussion. When prominent authorities such as Pinker or Summers make public statements suggesting that women have less natural aptitude for formal reasoning than men do, this contributes to the very environmental factors that prevent girls and women from fully realizing their aptitude—whatever it may be. Free discussion and open debate are important, but it is important as well to recognize the social effects that can result from debates between individuals who differ greatly in social status. The sharp response of the critics is not (as is often suggested) simply because they find statements like those Summers made offensive but because they believe such statements to have a cost—to be harmful to the flourishing of many people—and because they wish to respond in a way that carries enough social weight to counteract the public influence that such statements carry when made by individuals with the social authority of people like Summers or Pinker—to carry the debate forward on somewhat less unequal terms.

Feminists and other critics of existing social arrangements are especially concerned to correct the mistake of treating the deliberate use of social power—and its effects—as simply part of "nature." Consider an elementary example that evolutionary psychologists often cite—the human sweet tooth. The standard story goes as follows. Humans evolved to enjoy sweet foods because in ancestral environments such foods were relatively rare, and their sweetness was an indicator of the valuable caloric density that they supplied. In today's developed countries, however, food is plentiful and sugary foods are ubiquitous. In these circumstances, people have a natural tendency to consume far more sugary food than they need, with effects that are harmful to their health. But though this story is basically true, it is also a dramatic oversimplification. Humans are indeed attracted to sweet foods, yet this alone does not explain the extraordinary amount of refined sugar consumed by North Americans. People elsewhere in the world have not yet adopted such sugar-laden diets. The full explanation of sugary North American diets would require exploration of the place of sugar cane in nineteenth century colonial economies and its relation to slavery, of government subsidies for corn production and their effects on the market for corn syrup in the late twentieth century, and of the rise of the advertising industry and especially of television advertising aimed at children. The underlying human tendency to like sweet foods is involved in all these stories, but so are systematic and successful efforts to manipulate that tendency and to exaggerate its effects by entangling it with other evolved human responses such as the desire for social belonging. As this case begins to suggest, to mistake the effects of social manipulation for a mere expression of human nature is of particular consequence today since the new understanding of some channels for shaping human behavior that is now emerging from research in social and cognitive psychology is already being put into powerful use by private interests, affecting not only what food we choose but also many other aspects of our behavior, our preferences, our identities, and our values.

Chapter 3 compares evolutionary psychologists' inferences regarding human nature with inferences based on our understanding of the "natural" limits on human longevity but notes that recent research in the science of aging gives some reason to think that the limits to life expectancy might be overcome by interventions aimed at modifying the action of mechanisms of cell repair and aging—what might be called "internal interventions" since their effect is to modify the internal functioning of the organism directly. A similar possibility obviously arises for the kinds of features discussed under the rubric of "human nature." The last few chapters discuss interventions on social and environmental leverage points, but many internal interventions—especially those involving hormones that help regulate brain function—are now becoming much more accessible. We are already treating a large proportion of the North American population with hormones or drugs that mimic or modify the functioning of hormones—including hormones involved in shaping sex differences, social dominance, aggression, and empathy—in ways that sometimes reinforce or exaggerate features of "human nature" and sometimes aim to change them. Hormone treatments for developing fetuses are still a rarity but can be expected to become more common; it is noteworthy that one of the first uses developed is aimed at preventing "masculinization" in female fetuses. Such interventions early in the developmental process—in utero or in infancy—offer a set of levers by which existing social patterns could be reinforced, exaggerated, or changed. The limits of feasibility here are quite unknown, and vital questions of value are obviously inadequately explored to date.

Internal interventions work at the smallest scale to create social change. At the other extreme, interventions can also be aimed at affecting the functioning of entire self-sustaining behavior settings: modifying or reinforcing existing settings or creating novel ones. These interventions can be remarkably powerful in their effects: changing the boundary rules of educational settings so that women and people of color could occupy the role of university student, for example, had effects that have not ceased ramifying a century later. New technologies also contribute to the creation of novel behavior settings with far-reaching effects on people's development and behavior. The use of some kinds of behavioral leverage points has recently become the focus of vigorous public debate about the prospects and legitimacy of using the simple behavioral "nudges" we discussed earlier to help people make choices that accord better with their own rational self-interest (as economists understand it): to increase their retirement savings, reduce their debt burden, eat healthier food, or quit smoking, for example (Sunstein 2014; Thaler and Sunstein 2008). The rapid uptake by some governments of the idea of changing citizen's behavior by means of manipulation rather than coercion suggests that such methods, if effective, will be rapidly extended to modify behavior in ways intended to serve purposes beyond individual self-interest: to prevent crime, bullying, or littering or to reduce individual energy consumption, for example. Debate about these possibilities has focused on their implications for human welfare and freedom and has been informed mainly by economic theory and research in cognitive psychology (Sunstein 2014; Thaler and Sunstein 2008; White 2013). The broader array of evidence reviewed earlier, however, suggests that current discussions of the prospects and problems associated with behavioral nudges have barely scratched the surface of the issues that will need exploration as we learn more about the patterns of human response. Environmental adjustments may be able to take advantage of behavioral and developmental leverage points that not only change the degree to which people act on their long-term preferences despite short-term temptations—the classic nudge—but also change the preferences themselves, perhaps irreversibly and even heritably. Such interventions may have effects not just on what people do but in a real sense on who they are.

The ethical and political questions raised by such possibilities extend far beyond those now under discussion by proponents and critics of nudges as a means of carrying out a program of "libertarian paternalism." Can such interventions be morally acceptable? They seem like a frightening form of social engineering. Evolutionary psychologists warned that social engineering was difficult and could succeed only with difficulty or at a cost. But of course in certain regards easy and effective social engineering is even more alarming than social engineering that struggles against a resistant human nature. A government that claims the right to modify peoples' preferences, values, and identities seems to have overstepped its proper role, but all the more so if these aims can really be achieved. Yet thinking about the feasibility of achieving social goals by means of environmental adjustments also helps reveal the ways in which our current patterns of behavior are shaped or conditioned by the environments that we create for ourselves and for each other. Social engineering is a reality, whether we acknowledge it or not. To look for ways to make well-considered democratic choices about how to use its potentials, rather than leaving these choices up to private interests and undemocratic governments, is the morally responsible course.

[9]

Evolutionary Psychology and Human Possibilities

SOCIAL REFORMERS CONTINUE TO SEEK ways to make our societies more peaceful, inclusive, cooperative, democratic, and egalitarian, and the pressures that climate change, population growth, and other environmental constraints are now exerting on human life on earth give a new urgency to some of these concerns. The most prominent syntheses of evolutionary psychology, however, have concluded that evolved human nature stands in the way of at least some such "utopian" goals and warn that serious efforts to bring them about may do more harm than good. This book offers a critical study of these discussions and a survey of newer work on evolution, development, and behavior that combine to yield a more optimistic picture of a human nature capable of responsive change at both the individual and the social level. It also points to a new conception of both the scope of human evolutionary psychology and its role in discussions of social goals and policies.

The early chapters of this book examine two main bodies of literature. The first comprises the major popular syntheses of research and thinking about evolution and human behavior. These works have been influential in part because they address big issues of the limits of human social change, its possible goals, and the feasible mechanisms by which it can be realized. This valuable engagement with broad social issues and the success these writings have met in stimulating public discussion informed by scientific research make it important to understand both the weaknesses and the strengths of the claims they make. The second literature, drawn from evolutionary psychology, behavioral ecology, evolutionary and developmental biology, and social psychology, comprises more specialized studies of the processes of evolution, development, and behavior, including those involved in producing particular behaviors such as those important in gender relations, intergroup conflict, and social hierarchy. These researchers have usually been more interested in specific empirical questions than in the general social lessons that might be drawn from their research, although some have drawn general lessons about the need for reform in evolutionary psychology (Bolhuis et al. 2011). My aim has been to show how a philosophical critique of the more general studies and a close examination of the findings and thinking in the more specific studies taken together form the basis for a new understanding of evolutionary psychology that can make a vital contribution to discussion of the possibilities for human society.

The critique of the syntheses of evolutionary psychology by writers like Steven Pinker, Robert Wright, and Richard Dawkins shows that a proper understanding of the dynamics of evolution, development, and behavior does not support the conservative interactionism that they espouse. Their point of view is that the inbuilt qualities of human nature impose powerful restrictions upon the kinds of social change that are feasible and acceptable. But the arguments supporting this view make use of misleading metaphors, misapply cost–benefit thinking to the evaluation of change, and blur the distinction between fact and value in consequential ways—making assumptions about values that bias their judgments of matters of empirical fact and treating normative conclusions about what is good for humans and human societies as if they emerge straightforwardly from the facts of human evolutionary history.

If the critique calls into question the pessimistic conclusions of influential writing on evolutionary psychology, recently published research in evolutionary and developmental biology has revealed patterns of responsive change in behavioral capacities in humans and other species that require a very different conception of evolved human nature. The dynamics of evolution, development, and behavior that this research has uncovered indicate that there is a much more lively interaction between changes in the behavior of an organism and its environment than mainstream evolutionary psychology has assumed. The mechanisms of responsive change include niche construction and the environmentally cued "switches" characteristic of adaptive developmental and behavioral plasticity. Together these indicate that there are sometimes key points at which a small environmental intervention may trigger a distinct new process of sustained and accumulating change. They also suggest that change can sometimes be rapid and relatively smooth. It is as if there are leverage points where an environmental change across some specific threshold opens a new pathway for behavioral change that in turn has an impact on the environment. The new pathway may lead to further change, or it may arrive at a kind of stability or resilience.

The fields of study canvassed here have experienced rapid growth over the past few decades, and this new thinking continues to stimulate ongoing debate, but the older standard of conservative interactionism that posits human nature as a formidable obstacle to any efforts to diminish social and gender inequality and social conflict must be abandoned. The internalist bias of the synthesizers of evolutionary psychology has shielded them from the need to grapple with many of the issues raised by the possibility of choosing to pursue one or another path of change in human behavioral tendencies. Once that possibility is raised, however, it becomes important to gather the best understanding of the ways that interacting changes in human social arrangements and human environments work. There is good reason to look closely at the potential behavioral leverage points that research has already identified and to undertake new research with the aim of identifying other similar potential areas of intervention. Here the specific studies of evolution and behavior already provide suggestive clues and examples that give real hope for new understanding of possibilities of change. A prime example is Gowaty's idea that the sexual strategies of both women and men are the result of a flexible response to environmental conditions. Existing patterns are not hardwired at all; under specific developmental conditions the highly differentiated reproductive strategies now commonly attributed to female and male "natures" would be replaced by more similar strategies. If change is indeed a valued outcome, then research into the mechanisms of change is justified.

Only after the mechanisms of change are better understood can well-informed discussion of the moral and political aspects of pursuing change be undertaken. The truncated idea of cost–benefit reasoning that the synthesizers of evolutionary psychology employ is clearly insufficient. The values of freedom, justice, and equality become crucial but must be understood in ways that take account of the dynamic and diverse ways in which humans respond to their environments. The concept of "capability" helps to evaluate the ways in which people's preferences may be altered through the impact of selected environmental changes. With the recognition that the environment is changing in ways that require new human social patterns, the moral issues involved in creating a durable consistency in the relationship between human society and its environment can no longer be avoided. The possibility that social or political intervention might theoretically trigger important changes in behavior–environment relations raises the crucial question of feasibility. Does the research and thinking in evolutionary biology and evolutionary psychology suggest a preliminary set of feasible points of interaction that might lead to changes in activities that are stable and resilient in the face of the changing climate of our planet? I have shown that the standpoint of evolutionary psychology, properly understood, together with recent research in the field give reason to believe that there are feasible pathways to the kinds of change that our changing environment requires.

The new thinking about mechanisms of change has begun to spark a wider discussion about its political and ethical implications. That discussion is necessarily in part a political one: the problems involved in weighing different distributions of diverse goods among individuals who differ in the values they hold—and whose values can change in response to actions undertaken by themselves or by others—are not technical problems but political ones, requiring real engagement and effective negotiation among people with diverse situations and perspectives. The political discussion needs to be informed by the fullest available understanding of human patterns of development and behavior and their broader ramifications. This means that it must be grounded in a grasp of the workings of evolved and evolutionary processes of change in human development, behavior, and social arrangements. But it will benefit too from a lively dialogue with older traditions of thought about political change that look far beyond a narrow and economistic cost–benefit analogy to consider issues of democratic control, responsible leadership, and sources of social satisfaction. In addition, the creative thinking we need today must take account of the complex of constraints and forces hidden in the phrase "climate change" as it impacts human life globally and locally; here too evolutionary thinking can usefully combine with other modes of inquiry to help provide the needed understanding.

The discussion through which we make collective judgments about what kinds of social change are feasible and what paths of change are good to pursue is crucial to our ability to respond effectively and morally to the challenges now confronting humanity. Evolutionary thinking about human behavior—evolutionary psychology properly understood—can be a useful and vital part of that discussion.

Notes

1. Human Nature and the Limits of Human Possibility

1. In the context of evolutionary psychology, the term *sex* is usually used narrowly: sex is a matter of gamete size.

2. The Cost of Change

1. Reading this final category of costs broadly, we can see it as encompassing some of the others. Happiness and at least the subjective feeling of freedom are themselves phenotypic traits that can be traded off against others. Because of the special roles that happiness and freedom play in ethical thought, however, it is more useful here to treat them separately.

3. Thinking About Change and Stability in Living Systems

1. For this proportionality of cause to effect to hold in general requires not only that the causal factors combine additively but also that they act linearly. In mathematical terms, additivity actually follows from linearity so the strong sense of "addition" of causal factors that I describe here really amounts to the presumption that the causal function of genotype and environmental variables that determines the phenotype is a linear one.

2. In the technical language of causal theory, this sort of addition of causes is not classified as interaction. Interaction in the technical sense occurs precisely when the causal factors combine nonadditively so that a genetic difference may play out differently in different environments; an

environmental difference may have different consequences for different genotypes. I've followed Godfrey-Smith's usage here in including addition under the rubric of interaction because doing so usefully emphasizes that addition of causes is a limiting case of a much broader range of possible ways in which the internal and external factors can combine.

3. Lewontin attributes this remark to Arturo Rosenblueth and Norbert Wiener (1945), but it does not appear there.

5. Human Possibilities

1. How rare the strategy will be depends on the payoffs associated with conflict. If the cost of conflict is high enough, hawks may be the rare variant. But for the kinds of values often assumed in these models, the dove strategy will persist only at a low frequency. The key point is that this way of explaining desirable traits that are now rare implies that they cannot be stably maintained at frequencies much higher than what we now observe.

References

Ackerman, F., and L. Heinzerling. 2004. *Priceless: On Knowing the Price of Everything and the Value of Nothing.* New York: New Press.

Agassi, J. B. 1989. "Theories of Gender Equality: Lessons from the Israeli Kibbutz." *Gender & Society* 3, no. 2: 160–86.

Ah-King, M., and S. Nylin. 2010. "Sex in an Evolutionary Perspective: Just Another Reaction Norm." *Evolutionary Biology* 37, no. 4: 234–46.

Alkire, S. 2002. *Valuing Freedoms: Sen's Capability Approach and Poverty Reduction.* Oxford: Oxford University Press.

Amin, T., and N. S. Thompson. 2001. "Evolutionary Psychology Returns to Its Bowlbian Roots." *Behavior and Philosophy* 29:79–93.

Anderson, W. W., Y. K. Kim, and P. A. Gowaty. 2007. "Experimental Constraints on Mate Preferences in Drosophila Pseudoobscura Decrease Offspring Viability and Fitness of Mated Pairs." *Proceedings of the National Academy of Sciences of the United States of America* 104, no. 11: 4484–88.

Banzhaf, H. S. 2009. "Objective or Multi-Objective? Two Historically Competing Visions for Benefit–Cost Analysis." *Land Economics* 85, no. 1: 3–23.

Barker, G. 2008. "Biological Levers and Extended Adaptationism." *Biology & Philosophy* 23, no. 1: 1–25.

Barker, R. G. 1963. "The Nature of the Environment." *Journal of Social Issues* 19, no. 4: 17–38.

———. 1968. *Ecological Psychology: Concepts and Methods for Studying the Environment of Human Behavior.* Stanford, CA: Stanford University Press.

Baron-Cohen, S. 2002. "The Extreme Male Brain Theory of Autism." *Trends in Cognitive Sciences* 6, no. 6: 248–54.
Bateson, P. P. G. 1983. *Mate Choice*. Cambridge: Cambridge University Press.
Bateson, P. P. G., and P. D. Gluckman. 2011. *Plasticity, Robustness, Development and Evolution*. Cambridge: Cambridge University Press.
Bateson, P. P. G., and P. Martin. 1999. *Design for a Life: How Behaviour Develops*. London: J. Cape.
———. 2013. *Play, Playfulness, Creativity and Innovation*. Cambridge: Cambridge University Press.
Bial, D., and A. Rodriguez. 2007. "Identifying a Diverse Student Body: Selective College Admissions and Alternative Approaches." *New Directions for Student Services* 2007, no. 118: 17–30.
Boix, C., and F. Rosenbluth. 2007. "Bones of Contention: The Political Economy of Height Inequality." *American Political Science Review* 108, no. 1: 1–22.
Bolhuis, J. J., G. R. Brown, R. C. Richardson, and K. N. Laland. 2011. "Darwin in Mind: New Opportunities for Evolutionary Psychology." *PLOS Biology* 9, no. 7: e1001109.
Boomsma, J. J., and S. Nygaard. 2012. "Old Soldiers Never Die." *Genome Biology* 13, no. 2: 144.
Botero, C. A., and D. R. Rubenstein. 2012. "Fluctuating Environments, Sexual Selection and the Evolution of Flexible Mate Choice in Birds." *PLOS ONE* 7, no. 2: e32311.
Brizendine, L. 2006. *The Female Brain*. New York: Morgan Road Books.
Brown, G. R., T. E. Dickins, R. Sear, and K. N. Laland. 2011. "Evolutionary Accounts of Human Behavioural Diversity." *Philosophical Transactions B* 366, no. 1563: 313–24.
Browne, K. R. 2006. "Sex, Power, and Dominance: The Evolutionary Psychology of Sexual Harassment." *Managerial and Decision Economics* 27, no. 2–3: 145–58.
Buller, D. J. 2005. *Adapting Minds: Evolutionary Psychology and the Persistent Quest for Human Nature*. Cambridge: MIT Press.
Buss, D. M. 1995. "Psychological Sex Differences: Origins Through Sexual Selection." *American Psychologist* 50, no. 3: 164–71.
———. 2001. "Human Nature and Culture: An Evolutionary Psychological Perspective." *Journal of Personality* 69, no. 6: 955–78.

———. 2009. "How Can Evolutionary Psychology Successfully Explain Personality and Individual Differences?" *Perspectives on Psychological Science* 4, no. 4: 359–66.
Buss, D. M., and H. Greiling. 1999. "Adaptive Individual Differences." *Journal of Personality* 67, no. 2: 209–43.
Buss, D. M., and D. P. Schmitt. 1993. "Sexual Strategies Theory: An Evolutionary Perspective on Human Mating." *Psychological Review* 100, no. 2: 204.
———. 2011. "Evolutionary Psychology and Feminism." *Sex Roles* 64, no. 9–10: 768–87.
Buss, L. W. 1987. *The Evolution of Individuality*. Princeton, NJ: Princeton University Press.
Buss, T. F., and L. C. Yancer. 1999. "Cost–Benefit Analysis: A Normative Perspective." *Economic Development Quarterly* 13, no. 1: 29–37.
Chisholm, J. S. 1999. *Death, Hope, and Sex: Steps to an Evolutionary Ecology of Mind and Morality*. Cambridge: Cambridge University Press.
Churchland, P. S. 2011. *Braintrust: What Neuroscience Tells Us About Morality*. Princeton, NJ: Princeton University Press.
Colapinto, J. 2000. *As Nature Made Him: The Boy Who Was Raised as a Girl*. New York: HarperCollins.
Comim, F. 2005. "Capabilities and Happiness: Potential Synergies." *Review of Social Economy* 63, no. 2: 161–76.
Davis, L., and P. G. Stillman. 2005. *The New Utopian Politics of Ursula K. Le Guin's "The Dispossessed."* Lanham, MD: Lexington Books.
Dawkins, R. 1976. *The Selfish Gene*. Oxford: Oxford University Press.
———. 2006. *The God Delusion*. Boston: Houghton Mifflin.
Day, R. L., K. N. Laland, and F. J. Odling-Smee. 2003. "Rethinking Adaptation: The Niche-Construction Perspective." *Perspectives in Biology and Medicine* 46, no. 1: 80–95.
Dean, L. G., G. L. Vale, K. N. Laland, E. Flynn, and R. L. Kendal. 2014. "Human Cumulative Culture: A Comparative Perspective." *Biological Reviews of the Cambridge Philosophical Society* 89, no. 2 (May 14): 284–301.
Diamond, M., and H. K. Sigmundson. 1997. "Sex Reassignment at Birth: Long-Term Review and Clinical Implications." *Archives of Pediatric and Adolescent Medicine* 151, no. 3: 298–304.
Dupré, J. 1998. "Normal People." *Social Research* 65:221–48.

———. 2003. "On Human Nature." *Human Affairs* 13, no. 2 (December): 109–22.
Dweck, C. S. 2000. *Self-Theories: Their Role in Motivation, Personality, and Development*. Philadelphia: Psychology Press.
Eagly, A. H., and W. Wood. 2011. "Feminism and the Evolution of Sex Differences and Similarities." *Sex Roles* 64, no. 9–10: 758–67.
Ellis, B. J., M. Del Giudice, T. J. Dishion, A. J. Figueredo, P. Gray, V. Griskevicius, P. H. Hawley, W. J. Jacobs, J. James, and A. A. Volk. 2012. "The Evolutionary Basis of Risky Adolescent Behavior: Implications for Science, Policy, and Practice." *Developmental Psychology* 48, no. 3: 598.
Fairchild, H. H. 1991. "Scientific Racism: The Cloak of Objectivity." *Journal of Social Issues* 47, no. 3: 105–15.
Fehr, C. 2012. "Feminist Engagement with Evolutionary Psychology." *Hypatia* 27, no. 1: 50–72.
Fine, C. 2010. *Delusions of Gender: How Our Minds, Society, and Neurosexism Create Difference*. New York: W. W. Norton.
Forbes, C. E., C. L. Cox, T. Schmader, and L. Ryan. 2012. "Negative Stereotype Activation Alters Interaction Between Neural Correlates of Arousal, Inhibition and Cognitive Control." *Social Cognitive and Affective Neuroscience* 7, no. 7: 771–81.
Friedman, R. S., and B. Sutton. 2013. "Selling the War? System-Justifying Effects of Commercial Advertising on Civilian Casualty Tolerance." *Political Psychology* 34, no. 3: 351–67.
Gasper, D. 2005. "Subjective and Objective Well-Being in Relation to Economic Inputs: Puzzles and Responses." *Review of Social Economy* 63, no. 2: 177–206.
Gilbert, M. 2006. *The Disposable Male: Sex, Love and Money: Your World Through Darwin's Eyes*. Atlanta: Hunter Press.
Giovanola, B. 2005. "Personhood and Human Richness: Good and Well-Being in the Capability Approach and Beyond." *Review of Social Economy* 63, no. 2: 249–67.
Goodwin, B. C., and P. T. Saunders. 1992. *Theoretical Biology: Epigenetic and Evolutionary Order from Complex Systems*. Baltimore: Johns Hopkins University Press.
Gowaty, P. A. 1992. "Evolutionary Biology and Feminism." *Human Nature* 3, no. 3: 217–49.

———. 2003. "Power Asymmetries Between the Sexes, Mate Preferences, and Components of Fitness." In *Evolution, Gender, and Rape*, ed. C. B. Travis, 61–86. Cambridge: MIT Press.

———. 2008. "Reproductive Compensation." *Journal of Evolutionary Biology* 21, no. 5: 1189–1200.

———. 2011. "What Is Sexual Selection and the Short Herstory of Female Trait Variation." *Behavioral Ecology* 22, no. 6: 1146–47.

Gowaty, P. A., W. W. Anderson, C. K. Bluhm, L. C. Drickamer, Y. K. Kim, and A. J. Moore. 2007. "The Hypothesis of Reproductive Compensation and Its Assumptions About Mate Preferences and Offspring Viability." *Proceedings of the National Academy of Sciences of the United States of America* 104, no. 38: 15023–27.

Gowaty, P. A., and S. P. Hubbell. 2013. "Bayesian Animals Sense Ecological Constraints to Predict Fitness and Organize Individually Flexible Reproductive Decisions." *Behavioral and Brain Sciences* 36, no. 3: 215–16.

Halpern, D. F., L. Eliot, R. S. Bigler, R. A. Fabes, L. D. Hanish, J. Hyde, L. S. Liben, and C. L. Martin. 2011. "The Pseudoscience of Single-Sex Schooling." *Science* 333, no. 6050: 1706–7.

Hanlon, J., A. Barrientos, and D. Hulme, 2010. *Just Give Money to the Poor: The Development Revolution from the Global South*. Sterling, VA: Kumarian Press.

Hattenhauer, D. 1998. "The Politics of Kurt Vonnegut's 'Harrison Bergeron'." *Studies in Short Fiction* 35:387–92.

Hawks, J., E. T. Wang, G. M. Cochran, H. C. Harpending, and R. K. Moyzis. 2007. "Recent Acceleration of Human Adaptive Evolution." *Proceedings of the National Academy of Sciences* 104, no. 52: 20753–58.

Heft, H. 2001. *Ecological Psychology in Context: James Gibson, Roger Barker, and the Legacy of William James's Radical Empiricism*. Mahwah, NJ: Erlbaum.

Huxley, A. 1932. *Brave New World*. London: Chatto & Windus.

Kane, J. M., and J. E. Mertz. 2012. "Debunking Myths About Gender and Mathematics Performance." *Notices of the American Mathematical Society* 59, no. 1: 10–21.

Keller, J. 2007. "Stereotype Threat in Classroom Settings: The Interactive Effect of Domain Identification, Task Difficulty and Stereotype Threat on Female Students' Maths Performance." *British Journal of Education Psychology* 77, no. 2: 323–38.

Kelman, M. 2005. "Hedonic Psychology and the Ambiguities of 'Welfare'." *Philosophy & Public Affairs* 33, no. 4: 391–412.
Kitcher, P. 1985. *Vaulting Ambition: Sociobiology and the Quest for Human Nature*. Cambridge: MIT Press.
———. 1996. *The Lives to Come: The Genetic Revolution and Human Possibilities*. New York: Simon & Schuster.
———. 2003. *In Mendel's Mirror: Philosophical Reflections on Biology*. Oxford: Oxford University Press.
———. 2007. *Living with Darwin: Evolution, Design, and the Future of Faith*. Oxford: Oxford University Press.
———. 2011. *The Ethical Project*. Cambridge: Harvard University Press.
Kumm, J., K. N. Laland, and M. W. Feldman. 1994. "Gene-Culture Coevolution and Sex Ratios: the Effects of Infanticide, Sex-Selective Abortion, Sex Selection, and Sex-Biased Parental Investment on the Evolution of Sex Ratios." *Theoretical Population Biology* 46, no. 3: 249–78.
Lakoff, G., and M. Johnson. 1980. *Metaphors We Live By*. Chicago: University of Chicago Press.
Laland, K. N. 2006. "Animal Behaviour: Old World Monkeys Build New World Order." *Current Biology* 16, no. 8: R291–92.
Laland, K. N., J. Odling-Smee, and M. W. Feldman. 2000. "Niche Construction, Biological Evolution, and Cultural Change." *Behavioral and Brain Sciences* 23, no. 1: 131–75.
Le Guin, U. K. 1974. *The Dispossessed: An Ambiguous Utopia*. New York: Harper & Row.
Leger, D. W., A. C. Kamil, and J. A. French. 2001. "Fear and Loathing of Evolutionary Psychology in the Social Sciences." In *Nebraska Symposium on Motivation*, vol. 47, ed. D. W. Ledger, A. C. Kamil, and J. A. French, ix–xxiii. Lincoln: University of Nebraska Press.
Levy, B. R. 2003. "Mind Matters: Cognitive and Physical Effects of Aging Self-stereotypes." *The Journals of Gerontology Series B: Psychological Sciences and Social Sciences* 58, no. 4: P203–11.
Lewontin, R. C. 1974. "The Analysis of Variance and the Analysis of Causes." *American Journal of Human Genetics* 26, no. 3: 400–411.
———. 1996. "Evolution as Engineering." In *Integrative Approaches to Molecular Biology*, ed. J. Collado-Vides, B. Magasanik, and T. F. Smith, 1–10. Cambridge: MIT Press.
———. 2002. *The Triple Helix: Gene, Organism, and Environment*. Cambridge: Harvard University Press.

Liesen, L. T. 2007. "Women, Behavior, and Evolution." *Politics and the Life Sciences* 26, no. 1: 51–70.

———. 2011. "Feminists, Fear Not Evolutionary Theory, but Remain Very Cautious of Evolutionary Psychology." *Sex Roles* 64, no. 9–10: 748–50.

Lippa, R. A. 2009. "Sex Differences in Sex Drive, Sociosexuality, and Height Across 53 Nations: Testing Evolutionary and Social Structural Theories." *Archives of Sexual Behavior* 38, no. 5: 631–51.

Mani, A., S. Mullainathan, E. Shafir, and J. Zhao. 2013. "Poverty Impedes Cognitive Function." *Science* 341, no. 6149: 976–80.

McCarthy, M. M., and A. P. Arnold. 2011. "Reframing Sexual Differentiation of the Brain." *Nature Neuroscience* 14, no. 6: 677–83.

Mesoudi, A., and K. N. Laland. 2007. "Culturally Transmitted Paternity Beliefs and the Evolution of Human Mating Behaviour." *Proceedings of the Royal Society, Biological Sciences* 274, no. 1615: 1273–78.

Moore, A. J., P. A. Gowaty, and P. J. Moore. 2003. "Females Avoid Manipulative Males and Live Longer." *Journal of Evolutionary Biology* 16, no. 3: 523–30.

Mulder, M. B. 2004. "Are Men and Women Really So Different?" *Trends in Ecology & Evolution* 19, no. 1: 3–6.

Nelson, J. A. 2014. "The Power of Stereotyping and Confirmation Bias to Overwhelm Accurate Assessment: The Case of Economics, Gender, and Risk Aversion." *Journal of Economic Methodology* 21, no. 3: 211–31.

Nettle, D., D. A. Coall, and T. E. Dickins. 2011. "Early-Life Conditions and Age at First Pregnancy in British Women." *Proceedings of the Royal Society, Biological Sciences* 278: 1721–27.

Nosek, B. A., M. R. Banaji, and A. G. Greenwald. 2002. "Math = Male, Me = Female, Therefore Math Not = Me." *Journal of Personality and Social Psychology* 83, no. 1: 44–59.

Nosek, B. A., F. L. Smyth, N. Sriram, N. M. Lindner, T. Devos, A. Ayala, Y. Bar-Anan, et al. 2009. "National Differences in Gender-Science Stereotypes Predict National Sex Differences in Science and Math Achievement." *Proceedings of the National Academy of Sciences of the United States of America* 106, no. 26: 10593–97.

Nussbaum, M. C. 1995. *Human Capabilities, Female Human Beings: Women, Culture, and Development.* Oxford: Oxford University Press.

Odling-Smee, F. J., K. N. Laland, and M. W. Feldman. 2003. *Niche Construction: The Neglected Process in Evolution.* Princeton, NJ: Princeton University Press.

Orwell, G. 1949. *1984*. New York: Harcourt Brace Jovanovich.
Oyama, S. 1985. *The Ontogeny of Information: Developmental Systems and Evolution*. Cambridge: Cambridge University Press.
———. 2000. *Evolution's Eye: A Systems View of the Biology-Culture Divide*. Durham, NC: Duke University Press.
Oyama, S., P. Griffiths, and R. D. Gray. 2001. *Cycles of Contingency: Developmental Systems and Evolution*. Cambridge: MIT Press.
Paluck, E. L. 2009. "Reducing Intergroup Prejudice and Conflict Using the Media: A Field Experiment in Rwanda." *Journal of Personality and Social Psychology* 96, no. 3: 574–87.
Paluck, E. L. 2011. "Peer Pressure Against Prejudice: A High School Field Experiment Examining Social Network Change." *Journal of Experimental Social Psychology* 47, no. 2: 350–58.
Paluck, E. L., and D. P. Green. 2009. "Prejudice Reduction: What Works? A Review and Assessment of Research and Practice." *Annual Review of Psychology* 60: 339–67.
Panksepp, J., and J. B. Panksepp. 2000. "The Seven Sins of Evolutionary Psychology." *Evolution and Cognition* 6, no. 2: 108–31.
Penke, L. 2010. "Bridging the Gap Between Modern Evolutionary Psychology and the Study of Individual Differences." In *The Evolution of Personality and Individual Differences*, ed. D. M. Buss and P. H. Hawley, 243–79. New York: Oxford University Press.
Perilloux, C., D. M. Lewis, C. D. Goetz, D. S. Fleischman, J. A. Easton, J. C. Confer, and D. M. Buss. 2010. "Trade-Offs, Individual Differences, and Misunderstandings About Evolutionary Psychology." *American Psychologist* 65, no. 9: 930–32.
Piff, P. K. 2014. "Wealth and the Inflated Self: Class, Entitlement, and Narcissism." *Personality and Social Psychology Bulletin* 40, no. 1: 34–43.
Piff, P. K., P. Dietze, M. Feinberg, D. M. Stancato, and D. Keltner. 2015. "Awe, the Small Self, and Prosocial Behavior." *Journal of Personality and Social Psychology* 108, no. 6: 883–99.
Piff, P. K., M. W. Kraus, S. Côté, B. H. Cheng, B. H. and D. Keltner. 2010. "Having Less, Giving More: The Influence of Social Class on Prosocial Behavior." *Journal of Personality and Social Psychology* 99, no. 5: 771–84.
Piff, P. K., D. M. Stancato, S. Côté, R. Mendoza-Denton, and D. Keltner. 2012. "Higher Social Class Predicts Increased Unethical Behavior." *Proceedings of the National Academy of Sciences* 109, no. 11: 4086–91.

Pigliucci, M. 2001. *Phenotypic Plasticity: Beyond Nature and Nurture.* Baltimore: Johns Hopkins University Press.

———. 2005. "Evolution of Phenotypic Plasticity: Where Are We Going Now?" *Trends in Ecology & Evolution* 20, no. 9: 481–86.

Pigliucci, M., C. J. Murren, and C. D. Schlichting. 2006. "Phenotypic Plasticity and Evolution by Genetic Assimilation." *Journal of Experimental Biology* 209, no. 12: 2362–67.

Pinker, S. 2002. *The Blank Slate: The Modern Denial of Human Nature.* New York: Viking.

———. 2005. "Psychoanalysis Q-and-A." *Harvard Crimson*, January 19, 2015. http://www.thecrimson.com/article/2005/1/19/psychoanalysis-q-and-a-steven-pinker-in-an/.

———. 2011. *The Better Angels of Our Nature: Why Violence Has Declined.* New York: Viking.

Rajakumar, R., D. San Mauro, M. B. Dijkstra, M. H. Huang, D. E. Wheeler, F. Hiou-Tim, A. Khila, M. Cournoyea, and E. Abouheif. 2012. "Ancestral Developmental Potential Facilitates Parallel Evolution in Ants." *Science* 335, no. 6064: 79–82.

Rendell, L., L. Fogarty, and K. N. Laland. 2011. "Runaway Cultural Niche Construction." *Philosophical Transactions B* 366, no. 1566: 823–35.

Richardson, R. C. 2007. *Evolutionary Psychology as Maladapted Psychology.* Cambridge: MIT Press.

Roberts, S. C., and A. C. Little. 2008. "Good Genes, Complementary Genes and Human Mate Preferences." *Genetica* 134, no. 1: 31–43.

Robertson, D. S. 1991. "Feedback Theory and Darwinian Evolution." *Journal of Theoretical Biology* 152, no. 4: 469–84.

Rosenblueth, A., and N. Wiener. 1945. "The Role of Models in Science." *Philosophy of Science* 12, no. 4: 316–21.

Ruse, M. 2012. *The Philosophy of Human Evolution.* Cambridge: Cambridge University Press.

Santos R. G., M. J. Chartier, J. C. Whalen, D. Chateau, and L. Boyd. 2011. "Effectiveness of School-Based Violence Prevention for Children and Youth: Cluster Randomized Controlled Field Trial of the Roots of Empathy Program with Replication and Three-Year Follow-Up." *Healthcare Quarterly* 14: 80–91.

Schmitt, D. P. 2005. "Sociosexuality from Argentina to Zimbabwe: A 48-Nation Study of Sex, Culture, and Strategies of Human Mating." *Behavioral and Brain Sciences* 28, no. 2: 247–74.

Schmitt, D. P., A. Realo, M. Voracek, and J. Allik. 2008. "Why Can't a Man Be More Like a Woman? Sex Differences in Big Five Personality Traits Across 55 Cultures." *Journal of Personality and Social Psychology* 94, no. 1: 168.

Schoggen, P., R. G. Barker, and K. A. Fox. 1989. *Behavior Settings: a Revision and Extension of Roger G. Barker's Ecological Psychology*. Stanford, CA: Stanford University Press.

Sen, A. 1992. *Inequality Reexamined*. Cambridge: Harvard University Press.

Sen, A., and M. Nussbaum. 1993. "Capability and Well-Being." *Quality of Life* 1, no. 9: 30–54.

Simpson, G. G. 1969. *Biology and Man*. New York: Harcourt.

Steele, C. M. 1997. "A Threat in the Air: How Stereotypes Shape Intellectual Identity and Performance." *American Psychologist* 52, no. 6: 613–29.

Sterelny, K. 2003. *Thought in a Hostile World: The Evolution of Human Cognition*. Malden, MA, Blackwell.

———. 2007. "Social Intelligence, Human Intelligence and Niche Construction." *Philosophical Transactions B* 362, no. 1480: 719–30.

———. 2011. "From Hominins to Humans: How Sapiens Became Behaviourally Modern." *Philosophical Transactions B* 366, no. 1566: 809–22.

———. 2012. *The Evolved Apprentice: How Evolution Made Humans Unique*. Cambridge: MIT Press.

Sunstein, C. R. 2014. *Why Nudge? The Politics of Libertarian Paternalism*. New Haven, CT: Yale University Press.

Teschl, M., and F. Comim. 2005. "Adaptive Preferences and Capabilities: Some Preliminary Conceptual Explorations." *Review of Social Economy* 63, no. 2: 229–47.

Thaler, R. H., and C. R. Sunstein. 2008. *Nudge: Improving Decisions About Health, Wealth, and Happiness*. New Haven, CT: Yale University Press.

Tomic, N., and V. B. Meyer-Rochow. 2011. "Atavisms: Medical, Genetic, and Evolutionary Implications." *Perspectives in Biology and Medicine* 54, no. 3: 332–53.

Tooby, J., and L. Cosmides. 1990. "On the Universality of Human Nature and the Uniqueness of the Individual: The Role of Genetics and Adaptation." *Journal of Personality* 58, no. 1: 17–67.

———. 1992. "The Psychological Foundations of Culture." In *The Adapted Mind: Evolutionary Psychology and the Generation of Culture*, ed. J. H. Barkow, L. Cosmides and J. Tooby, 19–136. New York: Oxford University Press.

Travis, C. B., ed. 2003. *Evolution, Gender and Rape*. Cambridge: MIT Press.

Valian, V. 1998. *Why So Slow? The Advancement of Women*. Cambridge: MIT Press.

van der Wal, A. J., H. M. Schade, L. Krabbendam, and M. van Vugt. 2013. "Do Natural Landscapes Reduce Future Discounting in Humans?" *Proceedings of the Royal Society B: Biological Sciences* 280, no. 1773: 2013-2295.

Varnum, M. E., C. Blais, R. S. Hampton, and G. A. Brewer. 2015. "Social Class Affects Neural Empathic Responses." *Culture and Brain*: 1–9.

Veenhoven, R. 2010. "Life Is Getting Better: Societal Evolution and Fit with Human Nature." *Social Indicators Research* 97, no. 1: 105–22.

Volk, A. A., J. A. Camilleri, A. V. Dane, and Z. A. Marini. 2012. "Is Adolescent Bullying an Evolutionary Adaptation?" *Aggressive Behavior* 38, no. 3: 222–38.

Vonnegut, K. 1961. "Harrison Bergeron." *Magazine of Fantasy and Science Fiction* 21, no. 4: 5–10.

Wagner, A. 2014. *Arrival of the Fittest: The Hidden Mechanism of Evolution*. New York: Current.

Wagner, G. P. 2014. *Homology, Genes, and Evolutionary Innovation*. Princeton, NJ: Princeton University Press.

Walia, I., H. S. Arora, E. A. Barker, R. M. Delgado 3rd, and O. H. Frazier. 2010. "Snake Heart: a Case of Atavism in a Human Being." *Texas Heart Institute Journal* 37, no. 6: 687–90.

Walton, G. M. 2014. "The New Science of Wise Psychological Interventions." *Current Directions in Psychological Science* 23, no. 1: 73–82.

Walton, G. M., and S. J. Spencer. 2009. "Latent Ability: Grades and Test Scores Systematically Underestimate the Intellectual Ability of Negatively Stereotyped Students." *Psychological Science* 20, no. 9: 1132–39.

Webster, G., and B. C. Goodwin. 1996. *Form and Transformation: Generative and Relational Principles in Biology*. Cambridge: Cambridge University Press.

West-Eberhard, M. J. 2003. *Developmental Plasticity and Evolution*. Oxford: Oxford University Press.

White, M. D. 2013. *The Manipulation of Choice: Ethics and Libertarian Paternalism*. New York: Palgrave Macmillan.

Whitman, D. W., and A. A. Agrawal. 2009. "What Is Phenotypic Plasticity and Why Is It Important?" In *Phenotypic Plasticity of Insects:*

Mechanisms and Consequences, ed. D. W. Whitman and T. N. Ananthakrishnan, 1–63. Enfield: Science Publishers.

Wilkinson, R. G., and K. Pickett. 2010. *The Spirit Level: Why Greater Equality Makes Societies Stronger*. New York: Bloomsbury Press.

Wilson, D. S., E. Dietrich, and A. B. Clark. 2003. "On the Inappropriate Use of the Naturalistic Fallacy in Evolutionary Psychology." *Biology & Philosophy* 18:669–82.

Wilson, D. S., S. C. Hayes, A. Biglan, and D. D. Embry. 2014. "Evolving the Future: Toward a Science of Intentional Change." *Behavioral and Brain Sciences* 37, no. 4: 395–416.

Wilson, E. O. 1975. *Sociobiology: the New Synthesis*. Cambridge: Belknap Press of Harvard University Press.

———. 1978. *On Human Nature*. Cambridge: Harvard University Press.

Wilson, M., and M. Daly. 1997. "Life Expectancy, Economic Inequality, Homicide, and Reproductive Timing in Chicago Neighbourhoods." *British Medical Journal* 314, no. 7089: 1271–74.

Wright, R. 1994a. "Feminists, Meet Mr. Darwin." *New Republic* 211, no. 22: 34–46.

———. 1994b. *The Moral Animal: the New Science of Evolutionary Psychology*. New York, Pantheon Books.

Zhang, J. W., P. K. Piff, R. Iyer, S. Koleva, and D. Keltner. 2014. "An Occasion for Unselfing: Beautiful Nature Leads to Prosociality." *Journal of Environmental Psychology* 37: 61–72.

Index

f denotes figure

abnormality, 72
achievability (of social change), 115
active plasticity, 4, 54–55, 59, 88
adaptive plasticity, 55, 56–57
adaptive preferences, 97–98, 104
adaptive responses, 88, 100
adaptive strategies, 76, 88
additivity/addition, 38–39, 40f–41f, 44, 45, 141n
Anthem (Rand), 109
anticipatory plasticity, 57
Aristotle, 31, 32
autism, 32
averages, population/trait, 70, 92, 103

Barker, Roger G., 53
Bateson, Patrick, 11
behavior: niche construction and, 75; sexual differentiation in, 10, 15, 18–19, 36; as trick of rapid movement, 56

behavior–environment relations, intervention and changes in, 138
behavior settings, 65–66, 131–32
The Better Angels of Our Nature: Why Violence Has Declined (Pinker), 125
biofatalism, 3
biological levers, 60
biotic niche construction, 63
blank slate, 39, 61–62
Brave New World (Huxley), 101
bullying, 88, 100, 124, 132
Buss, David, ix, 5–6, 61, 112

capability, concept of, 138
capability approach, 98, 104–105
change, social: benefits of, 91; costs of, 22–34, 45, 90–91, 93, 107, 109–110; valuing of, 85–96
Christian tradition, on human nature, 31
Churchland, Patricia, 69

cognition, sexual differentiation in, 10, 15, 19, 36
communes/communalism, 76, 87, 93, 94, 112
conditional strategies, 75, 78
conservative interaction/interactionism, 45f–46f, 51, 67–69, 86, 99, 115, 128
continuous variation, 64
Cosmides, Leda, 13, 71
cost–benefit analysis, 21, 22, 27, 28–31, 33, 90, 91–92, 93, 96, 106, 110, 138

Darwin, Charles, 11, 12, 24
date rape, 18
Dawkins, Richard, 5, 22–23, 43, 56, 87, 136
descriptive conceptions/premises, 7, 9, 11, 31, 72, 73
desirable traits, 142n
development: alternative possible trajectories, 105; atypical outcomes, 73
developmental program, 43, 49
developmental psychopathology model, 99
direct impression, 42, 54
discontinuous plasticity, 69
discontinuous variation, 55, 64–65
Disney Princess, 119
The Dispossessed: An Ambiguous Utopia (Le Guin), 102
divergent evolution, 73
diversity, 71–76
Dweck, Carol, 122
dystopian fiction, 26, 101

ecological inheritance, 62, 64
ecological psychology, 53, 65
economic inequality, 7
effort, as cost attending social change, 2, 23–24, 25, 30, 90, 109–110, 112
environmental determinism, 38, 40f–41f
environment of evolutionary adaptedness (EEA), 12, 74
environments: choosing of, 97–114; modifying of, 62, 67, 73, 75–76, 88, 118–19, 126, 131 (*see also* niche construction); response to, 45, 49–50, 53–60, 62–64, 68–69, 75–76, 79–83, 88–89, 102–104, 125, 138–39 (*see also* plasticity; response functions; response perspective); roles of, 50
equality: excessive striving for, 91; form and degree of as good social goal, 109; gender equality, 10, 82, 108; of intellectual performance, 125; of outcomes, 25, 26, 45f, 108; social equality, 65, 108; value of, 107–108, 138
essentialist approach, 32
evaluation (of social change), 89–92
evolutionary approach/thinking, ix, x, 3
evolutionary psychology: basic theoretical framework of, 13; contributions of, 4; on costs of social change, 21–22; criticism of by feminists and others, 128; on freedom, 95–96, 97, 103–104; on happiness, 93–95, 97; on

human nature, 15, 70, 86; and human possibilities, 2, 4, 135–40; implications of for social policy, 5; on individual variation, 71; on interaction between organism and environment, 86; on plasticity, 61; "pop" evolutionary psychology, 5; on prospects for social change, 34; relationship between feminism and, 6
external causation, 40f–41f
externalist view/externalism, 38, 39, 42, 44, 67, 80, 89, 91, 93

facultative response model, 80–81
feasibility (of social change), 115–33
feedback loops/processes/systems, 53, 54, 62–63, 76, 121, 122, 123
Fehr, Carla, 6
female autonomy, reproductive/sexual, 77, 79–81, 83, 99, 102
female brains, 10, 32, 129
female characteristics, 77
female "nature," 138
female reproductive success, 78, 79
feminist evolutionary psychology, 6
Finland, education system of, 108
fixed mindset, 122
freedom: evolutionary psychology on, 95–96, 97, 103–104; loss of, as cost attending social change, 90
freely chosen expenditures, 112

gametes, size of as basic difference between sexes, 15–16
gender equality, 10, 82, 108
gender inequality, 8, 137
gender stereotypes, 10, 94–98, 107
gene-level selection, 13
genes, roles of, 50
genetic determinism, 3, 38, 40f–41f
genotypes: causal function of, 141n; inequality between, 45f; and phenotypic plasticity, 54; response functions for, 37, 40f–41f, 45f–46f, 48f–49f
Godfrey-Smith, Peter, 37, 141–42n
"good" characteristics, 105
Gowaty, Patricia, 6, 76, 78–82, 98, 102, 138
growth mindset, 122

happiness: costs to, 94–95; evolutionary psychology on, 93–95, 97; freedom's relationship to, 104; a phenotypic trait, 27n; role of environments in determining, 101
hardwiring, 39, 43, 61–62, 79, 129
"Harrison Bergeron" (Vonnegut), 26, 108–109
Hobbes, Thomas, 2, 31, 111, 126
Homo (genus), 12
homosexuality, 72, 94
hormone treatments, 131
human evolution, pace of, 75
human nature: Christian tradition on, 31; efforts to overcome, 33; evolutionary psychology on, 15, 70, 86; idea of as problematic, 31; limits set by, 21, 67, 128; representations of, 12–20; scientists on, 31; social

human nature (*continued*)
institutions as expressing or restricting?, 111
human possibilities: discussions about, 70; evolutionary approaches to, 3; evolutionary psychology and, 2, 4, 135–40; Gowaty's picture of, 80; optimism about, 71, 110; range of, 37, 62; scope of, 2
human response functions: implications of plasticity for thinking about, 60; and longevity, 36; and sex-differentiated cognitive capacities and behaviors, 36; shapes of, 37, 69
Hume, David, 8
Huxley, Aldous, 101, 102

imposed expenditures, 112
impossible dreams, pursuit of, 26
impression: direct impression, 42, 54; metaphor of, 45, 50; susceptibility to, 62
impression model, 80
individual variation, 61, 70, 71, 73, 92
inequality: between genotypes, 45f; reductions in, 3; social inequality, 94; wealth inequality, 123
Inequality Reexamined (Sen), 109
interaction/interactionism, 38, 44, 137, 141–42n. *See also* conservative interaction/interactionism; radical interaction/interactionism

intercultural antagonism, 14
internal causation, 40f–41f, 45
internal interventions, 131
internalist bias, 137
internalist view/internalism, 37–38, 39–40, 42, 43, 44, 65, 67, 89, 91, 93, 112, 115, 128
interventions: cautions with, 128; as changing phenotypes of whole population, 106; costs of, 25, 29–30; determining effectiveness of, 116; direct social intervention, 79; to disarm stereotype threat, 123–24; for homelessness, 123; internal interventions, 131; kinds of, 29–30, 90; large effects of some environmental interventions, 21, 51; moral acceptability of, 133; to reduce sex differences, 82; sex differences as exaggerated or created by environmental interventions, 65; small interventions compared to large interventions, 127, 128, 137; as triggering positive feedback cycles, 121; use of leverage points in identifying areas of, 138; wise interventions, 118, 122

kibbutzim, 76, 87, 93, 94, 112
Kitcher, Philip, 11, 78, 81

Lamarck, Jean-Baptiste, 38
latent capacities, 74
latent plasticity, 58
Le Guin, Ursula K., 102–103

Le Guinean conclusion, 102
leveling down, 91, 95, 107, 108
leverage points, 30, 116, 118, 127–28, 131–32, 137, 138
"Leviathan," 126
Lewontin, Richard, 42, 45, 49
libertarian paternalism, 133
Liesen, Laurette, 6
limited malleability, 68, 69, 86, 89, 93, 115
longevity, 36, 130–31
low-frequency genetic strategies, 72

Madonna–whore complex, 17–18
male brains, 10, 32, 129
male characteristics, 77
male "nature," 138
male reproductive success, 78–79
malleability, 39, 43, 54, 62, 69, 86, 89
masculinization, 32, 131
mathematical ability, 26, 120–21, 129
mating strategies, impact of on sex differences, 16–17, 18
men. *See* male
metaphors: of blank slate (*see* blank slate); of conservative interaction, 49; criticism of use of, 40; of direct impression, 42; for external influence, 42, 50; of genome-as-recipe, 44; of hardwiring (*see* hardwiring); of impression, 50; of interaction, 44, 49; of limited malleability (*see* limited malleability); for limits set by human nature, 21; of malleability (*see* malleability);

misleading use of, 136; of plasticity, 54; of playing out of programs, 49; price of as eternal vigilance, 42; of programming, 39; of recipe, 43; of response, 45, 50, 51, 54; of sorting, 42, 50; of unfolding, 43, 50, 68; views about interaction of biological causes expressed in, 39
microinequities, 127
misfiring, 72–73

natural goods, 107
naturalistic fallacy, 8, 87
natural outcomes/environments, 87, 112
natural phenotypes, 112
natural social systems, 112–13
nature, state of, 2
neuroplasticity, 56
niche construction, 4, 53, 62–66, 67, 75–76, 80, 88, 104, 111, 126, 137
1984 (Orwell), 25, 101, 109
nonadaptive plasticity, 55
nonmonetary goods, 28
normative conceptions/conclusions/implications, 7, 9, 22, 31, 43, 44, 49, 51, 72–73, 136
norms of reaction, 35, 40f–41f, 45f–46f, 48f–49f
nudges, 117, 132, 133
Nussbaum, Martha, 98, 101

Oakeshott, Michael, 25–26
On the Origin of Species (Darwin), 11

ontogeny, metaphor of, 50
optimism/optimists, 34, 71, 110
Orwell, George, 25, 101, 102, 109
Oyama, Susan, 49

passive plasticity, 54
perfect general equality, 108
personal: as biological, 82; as political, 82
phenotypes/phenotypic patterns, 31–32, 35, 36, 40f–41f
phenotypic outcomes, 32, 40f, 45f, 75, 88, 108
phenotypic plasticity, 54–58, 76
phenotypic stability, 58, 63–64
phenotypic trade-offs, 27, 91, 105
physical niche construction, 63, 65
Pinker, Steven, ix, x, 5, 6, 24, 25–26, 32, 33, 87, 91, 92, 107, 112, 125–26, 129, 136
plasticity: active plasticity, 4, 54, 55, 59, 88; adaptive plasticity, 55, 56–57; anticipatory plasticity, 57; discontinuous plasticity, 69; in general, 4, 53–62, 64, 67, 68, 70, 73, 74, 76, 137; latent plasticity, 58; neuroplasticity, 56; nonadaptive plasticity, 55; passive plasticity, 54; phenotypic plasticity, 54–58, 76
Posse Foundation, 124
potentials, evaluation of, 89
poverty, 94, 123
process structuralists, 38
programming, 39, 43

racism, 7, 8, 14, 34
radical interaction/interactionism, 45, 47, 48f–49f, 51, 67, 68, 80, 128
Rand, Ayn, 109
rape, 7, 18
rape culture, 83
reaction, norms of, 35, 40f–41f
recipe, metaphor of, 43
reinforcement, cost of, 23, 110, 112
reproductive success, 77, 78, 79, 92, 100
resources, allocation of, 110–11
response, metaphor of, 45, 50, 51, 54
response functions: in general, 35–37, 60, 68–69; representation of, 40f–41f, 45f–46f, 48f–49f
response model, 80–81, 82–83
response perspective, 88, 89, 98, 103, 114, 115, 116, 127, 128
responsive change, 137
responsive sensitivity, 68–69
risky behaviors, 99
robustness, 68, 113, 114
rock star, 18
Romantics, on human nature, 31
Roots of Empathy Project, 124
Rotman Institute of Philosophy (University of Western Ontario), xi
Rousseau, Jean-Jacques, 2, 31, 102, 111
Ruse, Michael, 24, 25, 32, 77

scaffolds, 58
schemas, 118
Schmitt, David, 61
Selfish Gene (Dawkins), 5

selfishness, Dawkins on, 22–23
self-sustaining systems, 113
Sen, Amartya, 98, 101, 108, 109
sensitivity, 60, 68–69, 105
sexual differentiation, in behavior and cognition, 10, 15, 18–19, 36
sexual double standard, 17
sexual infidelity, 17
sexual strategies, 76–83, 98, 138
Simpson, G. G., 11, 12
slut-shaming, 83
social arrangements, limitations of, 2
social change: achievability of, 115; aspirations for, 7; cost-benefit analysis of (see cost–benefit analysis); environment-driven, 85; evaluation of, 89–92; feasibility of, 115–33; likely occurrence of, 1; moral questions about, 5, 25; optimism about prospects for, 4; pessimism about prospects for, 3; prospects for according to evolutionary psychology, 107
social contagion, 124
social engineering, 133
social equality, 65, 108
social inequality, 94
social niche construction, 63, 65, 126
social possibilities/prospects, 3, 12, 20
sorting, metaphor of, 42, 50
species-typical outcomes/traits, 71, 103
stability, 37, 58, 59, 60, 62, 64, 66, 69, 75, 116

state of nature, 2
status quo ante, 28
stereotypes, 118–22
stereotype threat, 120–22, 123
stone-age minds, 87
Summers, Lawrence, 128, 129
sustainability: of current social arrangements, 116–18, 123; of social change, 115
sweet tooth, 13, 72, 129–30
switches, 55, 64, 69, 137

Tooby, John, 13, 71
trade-offs, 26–27, 35, 37, 91, 105
trait averages, 70. *See also* averages, population/trait
transgender, 94–95
transition, social, 24, 27, 28–29, 90, 110

unfolding, metaphor of, 43, 50, 68
unhappiness, as cost attending social change, 90
unnatural environments, 112
unnatural goods, 107
unnatural phenotypes, 112
unnatural preferences, 95
utility units, 27
utopia/utopians, 25, 33–34, 70, 76, 92, 101, 108, 112, 115, 135

value, problem of, 92–96
Vonnegut, Kurt, 26, 108–109

Walton, Gregory, 118
wealth inequality, 123
well-being, as cost attending social change, 23, 24

West-Eberhard, Mary Jane, 49
Wilson, E. O., 5, 23–24, 30, 32, 91, 110, 112
wise interventions, 118, 122

women. *See* female
Wright, Robert, ix, 5, 24, 25, 43, 86, 136

xenophobia, 14